284
Current Topics in Microbiology and Immunology

Springer
Berlin
Heidelberg
New York
Hong Kong
London
Milan
Paris
Tokyo

D. Harris (Ed.)

Mad Cow Disease and Related Spongiform Encephalopathies

With 34 Figures and 17 Tables

Springer

Professor David A. Harris, MD, PhD
Department of Cell Biology and Physiology
Washington University School of Medicine
660 South Euclid Ave
St. Louis, MO 63110
USA
e-mail: dharris@cellbio.wustl.edu

Cover illustration by M. Jeffrey and L. González showing disease specific PrP accumulation around a neuron in the brain of a BSE affected cow as revealed by the immunohistochemical technique.

ISSN 0070-217X
ISBN 3-540-20107-6
Springer-Verlag Berlin Heidelberg New York

Library of Congress Catalog Card Number 72-152360

Springer-Verlag is a part of Springer Science+Business Media
springeronline.com

Cover design: design & production GmbH, Heidelberg
Typesetting: Stürtz AG
Printed on acid-free paper – 27/3150 ag 5 4 3 2 1 0

Preface

Bovine spongiform encephalopathy (BSE) has become the most publicly recognizable example of a group of fatal neurodegenerative diseases caused by prions. Prions are an unprecedented class of infectious particles composed exclusively of protein without any associated nucleic acid. The prototypical prion disease of animals is scrapie, which has been known for over 200 years. Although prion disorders have long been of interest to the scientific, medical, and veterinary communities, the emergence of the BSE epidemic in the UK in the late 1980s and 1990s catapulted discussion of these diseases onto the front pages of newspapers and into the halls of government. The appearance of BSE in other European countries, Japan, and most recently Canada, as well as the spread of chronic wasting disease among deer and elk in North America, has intensified the need to understand and control these unusual disorders. The problem was moved forcefully into the realm of public health in 1996 with the emergence of variant Creutzfeldt–Jakob disease (vCJD), a human form of BSE. Whether the number of vCJD cases, presently less than 200, will increase further in the coming years, or whether the epidemic will be self-limited remains to be seen.

In this volume, leading authorities review key facets of BSE and related spongiform encephalopathies of animals and humans. The subjects that are covered include the molecular features of prions, the scope of the BSE epidemic in France, the pathogenesis of BSE and scrapie, international monitoring of BSE, vCJD (clinical, pathological, molecular, and epidemiological aspects), and chronic wasting disease of cervids. The work presented in these chapters provides an essential background for understanding a group of emerging infectious diseases that have come to assume a prominent place among both specialists and the public at large.

List of Contents

List of Contributors

(Their addresses can be found at the beginning of their respective chapters.)

Prion Protein and the Molecular Features of Transmissible Spongiform Encephalopathy Agents

J. R. Silveira · B. Caughey · G. S. Baron

Laboratory of Persistent Viral Diseases, NIAID, NIH, Rocky Mountain Laboratories, 903 S. 4th St., Hamilton, MT 59840, USA

E-mail: gbaron@niaid.nih.gov

Abstract Transmissible spongiform encephalopathy (TSE) diseases, or prion diseases, are neurodegenerative diseases found in a number of mammals, including man. Although they are generally rare, TSEs are always fatal, and as of yet there are no practical therapeutic avenues to slow the course of disease. The epidemic of bovine spongiform encephalopathy (BSE) in the UK greatly increased the awareness of TSE diseases. Although it appears that BSE has not spread to North America, chronic wasting disease (CWD), a TSE found in cervids, is causing significant concern. Despite decades of investigation, the exact nature of the infectious agent of the TSEs is still controversial. Although many questions remain, substantial efforts have been made to understand the molecular features of TSE agents, with the hope of enhancing diagnosis and treatment of disease, as well as understanding the fundamental nature of the infectious agent itself. This review summarizes the current understanding of these molecular features, focusing on the role of the prion protein (PrP^{C}) and its relationship to the disease-associated isoform (PrP^{Sc}).

1 Introduction

Transmissible spongiform encephalopathy (TSE) diseases, or prion diseases, are neurodegenerative diseases found in a number of mammals, including man. Although they are generally rare, TSEs are always fatal, and as of yet there are no practical therapeutic avenues to slow the course of disease. In the late 1980s and early 1990s, a large outbreak of bovine spongiform encephalopathy (BSE) in the UK greatly increased the awareness of TSE diseases. Since then, nearly 180,000 cases have been confirmed in the UK (DEFRA BSE information General statistics-GB), and 129 people have died due to a new form of human TSE, termed new variant Creutzfeldt–Jakob disease (vCJD) (UK Monthly CJD statistics). Data suggest that these occurrences are causally linked to BSE (Bruce et al. 1997; Hill et al. 1997). Although it appears that BSE has not spread to North America, chronic wasting disease (CWD), a TSE found in cervids, is causing significant concern (for a review see Williams and Miller 2002). No cases of human disease have been causally associated with CWD (Belay et al. 2001), but it is too early to conclude that humans are resistant to CWD. Furthermore, the presence of CWD in wild ruminants that exist on the same range as domestic livestock evokes questions about the potential transmission of CWD between these species, and the generation of new strains that might be hazardous to humans. Although many questions remain, substantial efforts have been made in understanding the molecular features of TSE agents, with the hope of enhancing diagnosis and treatment of disease, as well as understanding the fundamental nature of the infectious agent itself.

1.1 Nature of the TSE Agent

Early genetic studies in mice found that the *Sinc* gene was important in host susceptibility to TSEs (Dickinson et al. 1968). Subsequent work revealed that *Sinc* coded for the prion (PrP) protein, a host-derived protein often found accumulating in an abnormal, protease-resistant form in the central nervous system and lymphoreticular tissues during TSE disease. Substantial evidence now suggests that abnormal forms of PrP are involved in the transmission and pathogenesis of TSE disease (for a review see Chesebro 1999), and the prion hypothesis proposes that an abnormal form of PrP is in fact the TSE infectious agent (Prusiner 1998). Strong support for the importance of PrP in TSEs is derived from infectivity experiments performed with PrP knockout mice. These mice appear to be completely protected against TSE disease, and their ability to propagate infectivity was eliminated (Bueler et al. 1993). Nevertheless, the exact nature of the TSE agent remains to be determined as discussed below.

1.2 Nomenclature of PrP Isoforms

The accumulation of a number of designations for the various isoforms of PrP in the TSE literature has compounded the challenge of understanding these peculiar infectious agents. To simplify these designations as best we can, we will use the term PrP^{C} to refer to PrP in its normal structure and conformation and the term PrP^{sen} to refer generically to protease-sensitive forms of PrP, whether normal (i.e., PrP^{C}) or not (e.g., various recombinant forms). The terms PrP^{Sc}, PrP^{CJD}, PrP^{BSE}, etc., refer to abnormal forms of PrP associated with the particular TSE disease, and depending on the individual usage, may variably imply one or more of the following properties: infectiousness, pathogenicity, protease resistance or a specific conformational state. More generically and operationally, we refer collectively to protease-resistant forms of PrP as PrP^{res} and use this term unless referring specifically to the abnormal PrP forms associated with a particular TSE disease. However, the structures and properties of various abnormal PrP molecules are diverse, making the precise definitions and use of these terms problematic.

1.3 PrP Protein

The cDNAs of both hamster and mouse PrP encode for polypeptides of 254 amino acids (Locht et al. 1986; Basler et al. 1986). However, an N-terminal signal peptide of 22 amino acids is removed from these molecules during biosynthesis (Hope et al. 1986; Bolton et al. 1987; Turk et al. 1988), and an additional 23 amino acids are removed from the C termini of the proteins during glycosylphosphatidylinositol (GPI) addition at Ser231 (Stahl et al. 1990a), resulting in a mature PrP polypeptide of 210 residues. A single disulfide bond in PrP forms a loop (Turk et al. 1988), which contains two consensus sites for Asn-linked glycosylation at residues 181 and 197. Addition of glycans at these sites generates three main glycoforms: unglycosylated, partially glycosylated, and fully glycosylated PrP. High mannose glycans added to the protein in the endoplasmic reticulum are converted to complex or hybrid glycans in the Golgi apparatus (Caughey et al. 1989). Approximately 60 individual oligosaccharide chain variants have been found on PrP, indicating that the partially- and fully-glycosylated proteins consist of numerous minor glycoforms (Rudd et al. 1999; Stimson et al. 1999).

The secondary structure of PrP^{sen} has been analyzed by circular dichroism (CD) and Fourier-transform infrared spectroscopy, revealing a preponderance of α-helical content, but little β-sheet (Pan et al. 1993). Additional studies have employed high-resolution nuclear magnetic resonance (NMR) to analyze various recombinant PrP^{sen} molecules (Wuthrich and Riek 2001). NMR analysis of a mouse PrP^{sen} molecule comprised of residues 121–231 revealed three α–helices and a short section of two-stranded β-sheet structure (Riek et al. 1996). Further NMR analyses have used more complete PrP proteins, such as a hamster PrP molecule consisting of the protease-resistant residues (~90–231) found in PrP^{res} (Liu et al. 1999), and full length (residues ~23–231) recombinant PrP^{sen} proteins from both hamster (Donne et al. 1997) and mouse (Hornemann et al. 1997; Riek et al. 1997). These analyses indicate that the residues N-terminal to the 121–231 domain do not have a defined three-dimensional structure, but are instead highly flexible and disordered. A recent crystal structure of recombinant human prion protein (residues 90–231) revealed a dimeric form of PrP in which the disulfide bond was rearranged, and the C-terminal helix 3 was swapped between the dimeric interface, suggesting a possible mechanism for oligomerization (Knaus et al. 2001).

PrP^{sen} that is translocated to the surface of the cell is anchored to the plasma membrane through the GPI anchor, and the bulk of this PrP is phospholipase sensitive (Stahl et al. 1987; Caughey et al. 1989; Stahl et al. 1990b; Caughey et al. 1990). PrP^{sen} on the cell surface has a half-life of 3–6 h (Caughey et al. 1989; Borchelt et al. 1990; Caughey and Raymond 1991), and most PrP appears to be degraded in nonacidic compartments bound by cholesterol-rich membranes (Taraboulos et al. 1995). Studies on the endocytosis of PrP^{sen} indicate that it cycles between the cell surface and an endocytic compartment with a transit time of ~60 min (Shyng et al. 1993a). Both sulfated glycans (Shyng et al. 1995) and copper (Pauly and Harris 1998) have been shown to stimulate the endocytic process; however, the exact mechanism of the internalization, whether through clathrin-coated pits (Shyng et al. 1994; Shyng et al. 1995b; Sunyach et al. 2003) or other noncoated pit mechanisms (Kaneko et al. 1997a; Marella et al. 2002) has been controversial. Additionally, a small percentage of PrP^{sen} also appears to be released from the cell into the surrounding milieu (Caughey et al. 1988, 1989; Borchelt et al. 1990; Parizek et al. 2001).

1.4
PrP Polymorphisms in Bovine and Cervid Species

Polymorphisms in the PrP gene are associated with variability in a number of aspects of TSE disease, including relative susceptibility, clinical course, incubation times, and pathological lesion patterns. Recent analyses of the PrP gene from Rocky Mountain elk revealed that the sequence was highly conserved, with only one amino acid polymorphism detected in the population: Met to Leu at codon 132 (O'Rourke et al. 1999) (Fig. 1). This site corresponds to polymorphic codon 129 in the human PrP gene (Met to Val), a site at which homozygosity predisposes exposed individuals to some forms of CJD (Collinge et al. 1991; Palmer et al. 1991; Zeidler et al. 1997). In both free-ranging and farm-raised Rocky

	96	98	100	112	115	132	136	138	141	142	146	148	158	169	171	173	177	187	189	206	208	218	222	223	226	230	232	233
Elk	G	T	S	M	V	M/L	A	S	L	I	N	Y	Y	V	Q	N	T	V	Q	I	M	I	Q	R	E	Q	G	A
Deer	G/S	-	-	-	-	M	-	S/N	-	-	-	-	-	-	-	-	-	-	-	-	-	-	-	-	Q	-	-	-
Sheep	-	S	-	-	-	M	-	-	-	-	-	-	-	-	-	S	N	-	-	-	I	-	-	-	Q	-	-	-
	93	95	97	109	112	129	133	135	138	139	143	145	155	166	168	170	174	184	186	203	205	215	219	220	223	227	229	230
Human	-	-	-	-	M	M/V	-	-	I	-	S	-	H	M	E	S	N	I	-	V	-	-	E	-	Q	-	-	S
	93	95	108	120	123	140	144	146	149	150	154	156	166	177	179	181	185	195	197	214	216	226	230	231	234	238	240	241
Cattle	-	-	G	-	-	M	-	-	-	-	S	-	H	-	-	S	N	-	E	-	-	-	-	-	Q	-	-	-
	93	95	97	109	112	129	133	135	138	139	143	145	155	166	168	170	174	184	186	203	205	215	219	220	223	227	230	231
Hamster	-	-	N	-	M	M	-	-	M	M	-	W	N	-	-	-	N	I	-	-	I	T	-	K	Q	D	R	S
	92	94	96	108	111	128	132	134	137	138	142	144	154	165	167	169	173	183	185	202	204	214	218	219	222	226	229	230
Mouse	-	-	N	L	-	M	-	-	M	-	-	W	-	-	-	S	N	I	-	V	-	V	-	K	Q	D	R	S

Fig. 1 PrP amino acid sequence variations at residues that differ amongst Rocky Mountain elk (*Cervus elaphus nelsoni*), mule deer (*Odocoileus hemionus*), white-tailed deer (*Odocoileus virginianus*), sheep (ov-AQ), humans (hu-M or hu-V at residue 129), cattle (bo), hamsters, and mice. Residues that differ from elk in the sequence comprising the protease-resistant core of PrP^{res} are indicated (excluding insertions and deletions). All the known cervid, human and bovine polymorphisms are represented, but only one of 11 known sheep allelic forms is shown. Variations amongst cervids occur at residues 96, 132, 138 and 226 (*boxed*). In elk, these residues are either G, M, S and E (e-GMSE) or G, L, S and E (e-GLSE), respectively (O'Rourke et al. 1999). Corresponding residues G, M, S and Q (md/wd-GMSQ) and G, M, N and Q (md/wd-GMNQ) are found in both mule deer and white-tailed deer, while to date the residues S, M, S and Q (wd-SMSQ) have been found only in white-tailed deer. Corresponding residue numbers for each species are provided *above* the single letter amino acid codes. (Adapted from Raymond et al. 2000)

Mountain elk, homozygosity for Met at codon 132 was found to be over-represented in CWD-affected elk relative to unaffected control groups (O'Rourke et al. 1999). Two polymorphisms have been reported in white-tailed and mule deer, a glycine/serine polymorphism at residue 96, and a serine/asparagine polymorphism at residue 138 (Fig. 1). Although data are not available on the prevalence of the various polymorphism combinations in CWD-affected deer, cell-free conversion studies indicate that the various polymorphic PrP^{sen} species are converted to PrP^{res} at different efficiencies in some instances (see Fig. 6), and suggest differential susceptibilities to CWD infection in vivo (Raymond et al. 2000). The only polymorphism to affect the bovine PrP protein is a difference in the number of octapeptide repeat sequences (either five or six copies), and the frequency of this genotype is the same in both healthy cattle and those with BSE (Hunter et al. 1994). Doppel (Dpl), a prion-like protein derived from a gene downstream of the gene for PrP, has also been analyzed for polymorphisms in cattle. Although three polymorphisms were found in the Dpl coding region, none of them was significantly associated with BSE (Comincini et al. 2001).

1.5 Normal Function of PrP^C

The normal function of PrP is not immediately apparent. However, altered sleep patterns and circadian activity rhythms have been observed in mice devoid of PrP (Tobler et al. 1996, 1997), and a host of studies have suggested other roles for the protein. PrP's ability to bind copper (Hornshaw et al. 1995; Brown et al. 1997; Viles et al. 1999; Jackson et al. 2001), combined with observations of reduced copper levels in scrapie-infected mice (Thackray et al. 2002), and sporadic Creutzfeldt–Jakob disease (sCJD) subjects (Wong et al. 2001a) suggests a role for PrP in the regulation of copper. However, analyses of copper levels in the brains of transgenic mice with altered levels of PrP expression have produced disparate results (Brown et al. 1997; Waggoner et al. 2000), confounding the question of PrP's role in copper metabolism. PrP has also been shown to have superoxide dismutase (SOD) activity, dependent on the association of copper with the octapeptide repeat region (Brown et al. 1999). A number of proteins have been shown to interact with PrP, including laminin (Graner et al. 2000) and the 37-kDa laminin receptor precursor (Rieger et al. 1997) as well as proteins involved in neuronal signaling processes (Spielhaupter and Schatzl 2001). PrP has also been implicated in signal transduction in neuronal cells (Mouillet-Richard et al. 2000), and appears to transduce neuroprotective signals through interaction with stress-inducible protein 1 (Zanata et al. 2002; Chiarini et al. 2002). However, since PrP has been shown to have both pro-apoptotic (Paitel et al. 2002, 2003) as well as anti-apoptotic effects (Kuwahara et al. 1999; Bounhar et al. 2001), its ultimate role in cell viability is still in question. Additionally, studies suggest PrP plays a role in the activation of lymphocytes (Cashman et al. 1990; Li et al. 2001).

2 TSE Disease-Associated Changes in PrP

No consistent difference in covalent structure is known to distinguish PrP^C and PrP^{res}. However, the two isoforms can be readily discriminated in other ways (reviewed by Prusiner 1998; Weissmann 1999; Caughey 2001). In scrapie-infected neuroblastoma cells, PrP^{Sc} is made from mature PrP^{C} (Borchelt et al. 1990) after it reaches the plasma membrane (Caughey and Raymond 1991). Conversion likely occurs either on the cell surface and/or along an endocytic pathway to lysosomes (Caughey and Raymond 1991; Caughey et al. 1991a; Borchelt et al. 1992; Baron et al. 2002). In a cell culture model of infection, PrP^{Sc} has a much slower turnover rate than PrP^{C}. PrP^{C} is fully and rapidly digested by proteinase K (PK), whereas PK usually removes only roughly 67 of the ~210 total residues from the N terminus of most PrP^{res} molecules (Oesch et al. 1985; Hope et al. 1986, 1988). This difference in PK sensitivity is the most convenient and commonly used technique for distinguishing these two isoforms of PrP. However, PK-sensitive forms of PrP can copurify with PrP^{Sc} aggregates (Caughey et al. 1995, 1997). Such forms may be related to the PK-sensitive PrP molecules possessing hidden antibody epitopes, and hence non-native conformations that accumulate during the course of TSE disease (Safar et al. 1998). These disease-associated, PK-sensitive forms of PrP have been referred to as $sPrP^{Sc}$. PrP^{C} is generally soluble in mild detergents, whereas PrP^{res} is much less soluble and tends to assemble into amorphous aggregates or amyloid fibril-like structures (scrapie-associated fibrils or prion rods) (Merz et al. 1981; Diringer et al. 1983; Prusiner et al. 1983; Guiroy et al. 1991, 1993, 1994; Will et al. 1996). Although PrP^{res} is most readily detected in the central nervous system, spleen, and lymphoid tissues of infected animals, it has also been detected in muscle (Bosque et al. 2002) and placenta (Race et al. 1998; Tuo et al. 2002; Andreoletti et al. 2002). One group has reported the detection of a protease-resistant, noninfectious isoform of PrP in the urine of TSE-affected animals and humans, termed $UPrP^{Sc}$ (Shaked et al. 2001b).

2.1 Folding and Unfolding of PrP^{res}

Infrared, CD and X-ray diffraction analyses have shown that, in contrast to PrP^{C}, PrP^{Sc} is predominantly β-sheet with lower proportional α-helix content (Caughey et al. 1991b, 1998a; Pan et al. 1993; Safar et al. 1993a, 1993b; Nguyen et al. 1995). Therefore, a key event in the conversion of PrP^{C} to PrP^{res} is the transition of a portion of the α-helical and/or disordered secondary structures to β-sheet. Limitations of current technology have prevented high-resolution structural analyses of the noncrystalline PrP^{res} aggregates. Recent insight into PrP^{res} structure has come from electron crystallography studies of two-dimensional crystals of N-terminally truncated PrP^{Sc}, but the relationship of the aggregates analyzed to TSE infectivity in the preparation is uncertain (Wille et al. 2002). Consequently, further information about the structure of PrP^{Sc} has been sought with unfolding studies (Safar et al. 1993a, 1993b, 1994; Oesch et al. 1994; Caughey et al. 1995; Kocisko et al. 1996).

Initial studies showed that hamster PrP^{Sc} can be irreversibly disaggregated and unfolded with guanidine HCl (GdnHCl) or acid treatments. These studies concluded that PrP^{Sc} is fully monomerized at ~1.5 M GdnHCl and then unfolded at 2.5–3 M GdnHCl via a monomeric molten globule-like folding intermediate (Safar et al. 1993a, 1994). More recent studies obtained evidence for more structural heterogeneity in PrP^{Sc} aggregates and a somewhat different unfolding/disaggregation pathway (Caughey et al. 1995, 1997). When isolated from 263K scrapie-infected hamster brain tissue without exposure to PK, at least two types of PrP molecules can be observed in PrP^{Sc} aggregates. Roughly half of the molecules appear to be fully sensitive to PK digestion and can be readily dissociated from the PK-resistant core polymers with 2.5–3 M GdnHCl or PK treatment (Caughey et al. 1995, 1997; Callahan et al. 2001). These may correspond to the PK-sensitive, disease-specific PrP molecules later described by others as $sPrP^{Sc}$ (Safar et al. 1998). The remaining PrP^{res} aggregates have three distinguishable domains within the polymeric structure. The N-terminal residues 23–~89 are sensitive to PK cleavage in the absence of denaturant. A second ~3 kDa domain beginning at around residue 90 can be unfolded reversibly and rendered PK-sensitive by exposure to 2.5–3 M GdnHCl. Another C-terminal domain (~16 kDa in the aglycosyl structure) is the most resistant to PK as the GdnHCl concentration is increased. This domain becomes sensitive to PK at 3–4 M GdnHCl in an irreversible unfolding process that is accompanied by depolymerization and losses of both converting activity and scrapie infectivity (Caughey et al. 1997). At higher GdnHCl concentrations (5 M), PrP^{res} can be unfolded into a predominantly random-coil conformation (Callahan et al. 2001). Upon dilution from 5 M GdnHCl, PrP refolds into a conformation high in α-helix as measured by CD spectroscopy, similar to the normal cellular isoform of PrP^{C}, providing evidence that PrP^{Sc} can be induced to revert to a PrP^{C}-like conformation with a strong denaturant (Callahan et al. 2001).

To investigate associations between PrP^{Sc} conformation and TSE infectivity, many groups have attempted to reverse the loss of infectivity after denaturation of PrP^{Sc}. These studies have met with variable results. Copper ions are reported to assist the refolding of PrP^{Sc} after partial unfolding in GdnHCl and loss of TSE infectivity (McKenzie et al. 1998). However, other investigators have been unable to restore scrapie infectivity lost to treatments with urea, chaotropic salts, and SDS (Prusiner et al. 1993; Riesner et al. 1996; Post et al. 1998; Wille and Prusiner 1999). For instance, Riesner et al. showed that PK-treated PrP^{res} fibrils (prion rods) can be disrupted with SDS to form 10-nm spherical particles with high α-helical content and no PK-resistance or infectivity. Upon removal of the detergent or addition of acetonitrile, high β-sheet, PK-resistant aggregates were generated but these aggregates lacked infectivity (Riesner et al. 1996; Post et al. 1998). Aggregated and PK-resistant forms of PrP that have an amorphous ultrastructure and much reduced infectivity have also been generated after DMSO treatment of PrP^{Sc} (Shaked et al. 1999). More recent work has provided evidence of a requirement for nonprotein components of prion rods, especially heparan sulfate/glycosaminoglycans, in the reconstitution of infectivity from DMSO-solubilized PrP^{Sc} molecules (Shaked et al. 2001a). These and other studies (e.g. Hill et al. 1999; Xiong et al. 2001) with various PK-resistant forms of PrP generated in vitro using purified PrP^{sen}, indicate that PK-resistance and aggregation alone are not sufficient for PrP to be infectious.

Although harsh chemical treatments can denature PrP^{res} polymers in vitro and reduce infectivity, these treatments are not therapeutically applicable. Hence, it is important to know whether PrP^{Sc} formation is reversible under physiological conditions. One analysis has indicated that under nondenaturing conditions favorable for the conversion reaction (pH 6, 200 mM KCl, 0.6% sarkosyl), no detectable solubilization/monomerization of prewashed PrP^{Sc} aggregates occurred over several days (Callahan et al. 2001). Given the limits of the immunoblot detection system, the solubility of PrP^{Sc} under these conditions should be less than 2 nM. It will be of interest to determine whether other physiological conditions or factors such as chaperones, degradative enzymes or pharmaceutical agents might enhance PrP^{Sc} turnover in vivo as has been observed with other types of amyloid deposits (Chernoff et al. 1995; Patino et al. 1996; Rydh et al. 1998; Moriyama et al. 2000; Kryndushkin et al. 2002). In this regard, Soto and colleagues have identified synthetic peptides dubbed 'beta breakers' which tend to destabilize PrP^{res} and other amyloids (Soto et al. 2000; Sigurdsson et al. 2000). In addition, treatment with certain cationic lipopolyamines, branched polyamines, or anti-PrP antibodies can cure scrapie-infected cells of PrP^{Sc} and infectivity, suggesting the existence of endogenous cellular mechanisms for PrP^{Sc} degradation (Supattapone et al. 1999b, 2001; Winklhofer and Tatzelt 2000; Enari et al. 2001; Peretz et al. 2001b).

Any therapeutic approach to TSE diseases based on disruption of PrP^{Sc} aggregates must give consideration to the fact that the neurotoxic and/or infectious forms of PrP associated with these diseases remain to be defined. If PrP^{res} is the most important neuropathogenic substance in TSE pathogenesis, then destabilizing factors as described above may be of therapeutic benefit. On the other hand, if smaller PrP^{res} aggregates are more pathogenic than larger ones, or if an intermediate or byproduct of PrP^{res} formation is more toxic than PrP^{res} itself, then destabilization of PrP^{res} deposits in vivo may have a detrimental rather than beneficial effect (Masel et al. 1999; Caughey and Lansbury 2003). Finally, if infectious and neurotoxic forms of PrP are different entities, it will be critical to develop therapeutics that target both forms.

2.2
Formation of PrP^{res} by Deleted Forms of PrP

Many studies have investigated the effect of deletions of various portions of PrP^{C} to identify regions important for generation of PrP^{Sc}. In scrapie-infected mouse neuroblastoma cells, PrP^{Sc} formation occurs in cells expressing PrP lacking N-terminal residues 23–88 (Rogers et al. 1993). Further incremental N-terminal deletions revealed that removal of residues 34–94 and 34–113 significantly reduced cell-free conversion of PrP^{sen}, the latter mutant having an even lower conversion efficiency than 34–94 in addition to producing a conversion product with an altered conformation as measured by PK cleavage (Lawson et al. 2001). Consistent with these observations, transgenic mice expressing PrP^{C} lacking residues 32–93 remained susceptible to scrapie infection but with extended incubation times, absence of histopathology in the brain, and lower levels of accumulated PrP^{res} and infectivity titers (Flechsig et al. 2000). A 106-residue deletion mutant of PrP lacking residues 23–88 and 141–176 (Δ106) was also shown to be capable of supporting infection, disease propagation and the formation of PrP^{Sc}-like molecules when expressed in transgenic mice (Supattapone et al. 1999a). Thus, these deleted portions of the PrP molecule are not necessary for PrP's role in this model of TSE disease at least. However, expression of PrP lacking residues 114–121 blocked PrP^{Sc} formation (Holscher et al. 1998). Thus, these residues, which form a hydrophobic stretch in a region of the PrP^{C} molecule that undergoes major conformational change in its conversion to PrP^{res}, appear to be critical to the conversion process. This conclusion is supported further by the inhibition of PrP^{res} formation by synthetic peptides containing at least some of these residues (Chabry et al. 1998, 1999). Vorberg and co-workers found that deletion of regions of mouse PrP^{sen} corresponding to the first β-strand (127–130), the second β-strand (160–163), or the first α-helix (143–153) all significantly inhibited their ability to act as substrates for PrP^{res} formation both in scrapie-infected neuroblastoma cells and a cell-free assay system, implicating a role for these regions in conversion (Vorberg et al. 2001). A recent structural model of PrP^{res} aggregates proposes a parallel β-helical structure, suggesting that certain PrP^{sen} deletions, such as those in $\Delta 106$ PrP^{sen}, could be tolerated provided they correspond to a complete turn(s) of β-helix (Wille et al. 2002). This might provide an alternative explanation for the results of Holscher et al. (1998) and Vorberg and Priola (2002), though the aggregates used in the modeling study above have not yet been proven to be infectious.

2.3
Structural Diversity in Abnormal TSE-Associated PrP Molecules

The terms PrP^{res}, PrP^{Sc}, etc. are often used to describe the TSE disease-associated form of the PrP protein. However, it is important to point out that since disease-associated PrP is found in many different forms, a collection of conformational states and entities will fall under these terms. TSE-associated PrP molecules can vary in resistance to proteolysis (Bessen and Marsh 1994; Tagliavini et al. 1994; Bessen et al. 1995; Safar et al. 1998; Caughey et al. 1998a; Tzaban et al. 2002; Horiuchi et al. 2002), insolubility in detergents (Somerville et al. 1989; Bessen and Marsh 1994; Muramoto et al. 1996), secondary structure (Caughey et al. 1998a), glycoform ratios (Kascsak et al. 1986; Somerville and Ritchie 1990; Monari et al. 1994; Collinge et al. 1996; Somerville et al. 1997; Rudd et al. 1999; Hope et al. 1999; Race et al. 2002), exposure of epitopes and conformational stability in denaturants (Safar et al. 1998; Peretz et al. 2001a; Safar et al. 2002), multi-spectral ultraviolet fluorescence

(Rubenstein et al. 1998), ultrastructure (McKinley et al. 1991; Giaccone et al. 1992; Jeffrey et al. 1994) and membrane topology (Hegde et al. 1999). Complicating matters further is the fact noted above that PrP^{Sc} can be isolated as a complex of both partially PK-resistant and fully PK-sensitive molecules (Caughey et al. 1995). Although the PK-sensitive molecules are not required for infectivity and self-propagating activity (Caughey et al. 1997), they might play a role in neuropathogenesis (Horiuchi and Caughey 1999a). Finally, manipulations of various recombinant and mutant forms of PrP have generated numerous forms of PrP which appear to share some, but not all, of the properties of bona fide TSE-associated PrP isolated from infected tissues. A current challenge is to understand which abnormal states of PrP are relevant to various aspects of TSE transmission and/or pathogenesis. The requirements may be different for infectious versus neurotoxic forms of PrP. To account for these uncertainties, Charles Weissmann has proposed the use of the term PrP* to connote the putative infectious form of PrP, whatever it may be (Weissmann 1991).

3 Potential Mechanisms of PrP^{res} Formation

The most commonly considered theoretical models of PrP^{res} formation are the heterodimer model (Griffith 1967; Bolton and Bendheim 1988; Prusiner 1998) and the nucleation (seed)-dependent polymerization models (Griffith 1967; Gajdusek 1988; Jarrett and Lansbury 1993; Lansbury and Caughey 1995). The heterodimer model proposes that PrP^{res} exists in a stable monomeric state that can bind PrP^{C}, forming a heterodimer, and catalyze a conformational change in PrP^{C} to form a homodimer of PrP^{res}. The PrP^{res} homodimer then dissociates to give two PrP^{res} monomers. Fundamental aspects of this model are that PrP^{res} is more stable thermodynamically than PrP^{C}, conversion of PrP^{C} to PrP^{res} is rare unless catalyzed by a preexisting PrP^{res} template, and the PrP^{res} homodimer tends to dissociate into monomers. According to the model, this process also requires the assistance of a hypothetical, species-specific factor termed 'protein X'. In the nucleated polymerization model, oligomerization/polymerization of PrP is necessary to stabilize PrP^{res} sufficiently to allow its accumulation to biologically relevant levels. Spontaneous formation of nuclei or seeds of PrP^{res} is rare because of the weakness of monovalent interactions between PrP^{C} molecules and/or the rarity of the conformers that polymerize. However, once formed, oligomeric or polymeric seeds are stabilized by multivalent interactions (Jarrett and Lansbury 1993). The spontaneous formation of seeds might involve the formation of domain-swapped PrP dimers via swapping of helix 3 and rearrangement of the disulfide bond as has been observed in the crystal structure of recombinant PrP (Knaus et al. 2001). A new model of PrP^{res} formation proposed the involvement of intermolecular disulfide bonds in PrP^{res} formation (Welker et al. 2001), but subsequent studies from cell-free conversions suggest this is unlikely (Welker et al. 2002).

In their various permutations, the heterodimer and nucleated polymerization models can overlap. For instance, autocatalysis (or templating) of the conformational change in PrP^{C} by PrP^{res} may be a feature of either the heterodimer or nucleated polymerization models. However, nucleated polymerization could also occur if PrP^{C} rapidly interchanges between high α-helix and high β-sheet conformers with the latter being stabilized greatly by binding to a preexisting polymer of PrP^{res}. In either type of model, a metastable PrP^{C} folding intermediate might most favorably interact with PrP^{res} in the conversion reaction. Such an intermediate might resemble those generated from recombinant PrP^{sen} or brain PrP^{C} under acidic conditions (Swietnicki et al. 1997; Zhang et al. 1997; Hornemann and Glockshuber 1998; Jackson et al. 1999; Baskakov et al. 2002; Zou and Cashman 2002) similar to those of endosomes or lysosomes.

A critical question is what form(s) of PrP is responsible for initiation and propagation of PrP^{res} in TSE disease processes in vivo. The nucleated polymerization model predicts that active PrP^{res} seeds could range in size from the minimum stable oligomeric nucleus to large polymers. Consistent with this prediction is the frequent observation that cell-free converting activity (see below) and infectivity are associated with a wide size range of PrP^{res} aggregates, but not monomers (Prusiner et al. 1993; Hope 1994; Caughey et al. 1995, 1997; J. Silveira and B. Caughey, unpublished results). In contrast, heterodimer-type models propose that discrete monomers (or in related permutations, small discrete oligomers) are the active autocatalytic units. Aggregation of PrP^{res} would then be a side effect. A theoretical consideration of the kinetic consequences and likelihoods of the two models (Eigen 1996) concluded that a strict noncooperative heterodimer model is highly unlikely. However, if a small oligomer (e.g., trimer or tetramer) served as a template in a highly cooperative autocatalytic reaction, then the model becomes more plausible. In any case, it was suggested that aggregation of PrP^{res} is likely to be an important 'prerequisite of infection'. Candidate small oligomers of PrP have been observed consisting of anywhere from trimers to octamers, although they have not yet been proven to possess either infectivity or self-induced converting activity (Riesner et al. 1996; Wille et al. 2002; Baskakov et al. 2002). Mathematical modeling of the nucleated polymerization mechanism predicted that systems of short polymers should grow the fastest (Masel et al. 1999).

How then do these models address the situations of familial and sporadic human TSEs that are not clearly of infectious origin? In these instances, the formation of an initial template or seed might be a spontaneous and stochastic event that can be potentiated by specific PrP mutations (Jarrett and Lansbury 1993; Lansbury and Caughey 1995; Prusiner 1998; Chiesa et al. 1998, 2000; Ma and Lindquist 2002). On the other hand, in TSE diseases of infectious origin, transmission might be explained by acquisition of pre-formed PrP^{res} templates/seeds. Nonetheless, without a complete understanding of the nature of the TSE agent, it seems appropriate to consider the possibility that another type of agent, with or without a PrP component, is responsible for TSE transmission and the initiation of PrP^{res} formation in infected hosts.

4
Specific Binding of PrP^{sen} to PrP^{res} Aggregates

According to theoretical models, PrP^{res} and PrP^{C} are expected to interact directly in the course of TSE pathogenesis. Studies in animals (Prusiner et al. 1990), and scrapie-infected tissue culture cells (Priola et al. 1994; Priola and Chesebro 1995), provided the first empirical indications that PrP^{res} and PrP^{sen} may interact specifically. Specific binding of PrP^{res} to PrP^{sen} has been observed directly in lysates or culture supernatants from ^{35}S-methionine-labeled cells expressing soluble PrP^{sen} lacking the GPI anchor (Horiuchi and Caughey 1999b). Under these conditions (pH 6.0, 1% sarkosyl), only GPI^{-} ^{35}S-PrP^{sen} bound, out of the many labeled proteins in the preparations. Recent work exploring the interaction and conversion of membrane-bound PrP species did not detect conversion of membrane-anchored (GPI^{+}) PrP^{C} by exogenous PrP^{res} (Baron et al. 2002), suggesting that membrane attachment of PrP^{C} may limit accessibility to PrP^{res}. The potential source of the inaccessibility may be the location of PrP^{res} binding sites near the C terminus of PrP^{sen} (see below). These sites may be blocked by attachment to the membrane via the C-terminal GPI anchor (Horiuchi and Caughey 1999a), or alternatively, direct interactions between membrane surfaces and the PrP^{C} polypeptide chain may block its access to PrP^{Sc} (Morillas et al. 1999; Sanghera and Pinheiro 2002), although data from cell-free studies argue against the latter possibility (Baron and Caughey 2003).

5
PrP^{res}-Induced Conversion of PrP^{sen} to PrP^{res}

A key facet of the 'protein only' models for TSE disease is the concept that the disease-associated, abnormal form of PrP (PrP^{Sc}) possesses the ability to interact with the normal PrP protein (PrP^{C}) and 'convert' it into additional PrP^{Sc}. To date, a number of cell-free systems have been developed which promote the formation of new PrP^{res} from a pool of PrP^{C}. Although the generation of newly formed infectious particles remains to be demonstrated in an in vitro setting, these systems are capable of generating partially protease-resistant PrP (PrP^{res}) that is indistinguishable from TSE-associated PrP^{res} itself. While naturally lending additional support to the protein-only models of TSE disease, these cell-free systems also provide valuable tools for TSE research.

The available methods for cell-free PrP^{res} formation vary widely in both the purity of their components, as well as the levels of newly formed PrP^{res} obtained relative to the amount of seed PrP^{Sc} present. The simplest and most biochemically defined systems employ purified, ^{35}S-labeled PrP^{sen}, which is mixed with an excess of purified, unlabeled PrP^{res} (Kocisko et al. 1994). Upon incubation, the ^{35}S-labeled PrP^{sen} can first bind the PrP^{res} in a protease-sensitive state, then slowly convert to a protease-resistant state (DebBurman et al. 1997; Horiuchi and Caughey 1999b; Horiuchi et al. 2000). The newly formed PrP^{res} is detected as ^{35}S-labeled PrP species that show the characteristic partial resistance to digestion by PK. The conversion obtained in this system can be stimulated by a number of factors, including the addition of chaotropes, detergents, chaperone proteins, sulfated glycans, or heat (Kocisko et al. 1994; DebBurman et al. 1997; Horiuchi and Caughey 1999b; Wong et al. 2001b). These types of assays have also been adapted to a solid phase format, suitable for many applications including high-throughput screening of therapeutics (Maxson et al. 2003). More recently, this technique has been applied to membrane-bound forms of PrP^{sen} and PrP^{res} (Baron et al. 2002; Baron and Caughey 2003), and an in situ form of the assay has also been used to analyze the pattern of PrP^{res} formation in TSE-infected brain slices (Bessen et al. 1997). Additionally, cell extracts have served as a source of PrP^{C} for in vitro conversion assays (Saborio et al. 1999; Vorberg and Priola 2002). These systems have made use of a specific PrP epitope in the pool of PrP^{C} to discriminate between newly formed and input PrP^{res}. Cellular conversion systems have also been devised as models of cell-to-cell propagation of PrP^{Sc} and infection. In these systems, the formation of PrP^{res} in intact cells is promoted by the addition of PrP^{Sc}-containing brain extracts (Korth et al. 2000; Vorberg and Priola 2002).

Although many of the above systems enable the production of PrP^{res} under relatively defined conditions, the yield of newly formed PrP^{res} is usually less than that of the infectious PrP^{res} used to promote the reaction. Thus, the in vitro formation of 'converted' PrP that is infectious has been difficult to assess and remains to be demonstrated. However, new systems of in vitro conversion have been developed which substantially amplify PrP^{res} , and may finally enable the detection of newly formed, infectious PrP^{res}. The first of these techniques, known as protein misfolding cyclic amplification (PMCA) involves the combination of a small amount of detergent extract of TSE-infected brain homogenate with a vast excess of a similar extract of normal brain homogenate containing PrP^{C} (Saborio et al. 2001). Although the tissue homogenates used in this system are crude, the procedure, which involves repeated cycles of sonication and incubation, can elevate the detectable levels of PrP^{res} more than 30-fold. Compared to the modest levels of formation achieved in more purified systems, this represents a dramatic amplification of PrP^{res}. This difference has led to the suggestion that the brain homogenate may contain additional factors that catalyze conversion. Lucassen and colleagues have modified the PMCA technique, using only nondenaturing detergent, and removing the sonication step (Lucassen et al. 2003). Under these altered conditions, substantial amplification of PrP^{res} (>10-fold) was still obtained. Recently, another method of brain homogenate-based conversion has been reported (Zou and Cashman 2002). In this system, acidification and detergent treatment of PrP^{C} in normal brain homogenates promoted the formation of PrP^{Sc}-like, PK-resistant

PrP species. In the presence of trace quantities of PrP^{Sc} derived from CJD brain homogenates, this technique amplified the signal of PK-resistant PrP 3–10-fold. Either of these amplification techniques may provide the opportunity to test whether or not the newly formed PrP^{res} is infectious, however, the presence of crude brain extracts will still leave the question of whether factors besides PrP are critical in the composition and/or formation of TSE infectivity. Ultimately, it will be necessary to bridge the gap between biochemically defined PrP conversion reactions, which so far have not proven to generate new TSE infectivity, and the TSE agent propagation that is known to occur readily in intact TSE-infected cells and animals.

5.1 Biological Connections

Although cell-free generation of PrP^{res} has yet to demonstrate that it is an infectious protein, there are a number of ways in which even the simplest, most well defined systems of cell-free conversion reflect the biology of TSE disease in vivo. The PrP sequence specificity of the conversion reaction correlates with inter- and intra-species TSE transmissions in vivo, possibly reflecting an important control point in interspecies TSE transmissions (Kocisko et al. 1995; Raymond et al. 1997; Bossers et al. 1997, 2000; Raymond et al. 2000; Horiuchi et al. 2000). The strain specificity of the reaction mimics PrP^{res} formation in vivo and provides a potential mechanism for TSE strain propagation by a protein-only mechanism (Bessen et al. 1995). The correlation between cell-free converting activity and scrapie infectivity in GdnHCl denaturation studies suggests that these two parameters are related (Caughey et al. 1997). Finally, the fact that the in vitro conversion reaction has been adapted to physiologically compatible conditions suggests that it can also occur in vivo (DebBurman et al. 1997; Bessen et al. 1997; Horiuchi and Caughey 1999b; Wong et al. 2001b; Baron et al. 2002). Indeed, the in situ form of the conversion reaction revealed that both amyloid plaque and diffuse deposits of PrP^{res} have the ability to induce conversion (Bessen et al. 1997).

The ^{35}S-PrP^{res} product of the cell-free conversion reaction also mimics the behavior of PrP^{Sc} with regard to the reversibility of aggregation and domain structure (Callahan et al. 2001). PrP that binds to the PrP^{Sc} aggregate during the cell-free conversion reaction does not dissociate from the aggregate over the course of several days in conversion buffer lacking denaturant. Furthermore, when cell-free conversion reactions are digested with PK in the presence of 2.5 M GdnHCl, the resulting ^{35}S-PrP^{res} band was ~3 kDa smaller than the band resulting from digestion without GdnHCl. This partial unfolding of the PK-resistant core of newly converted PrP^{res} was at least partially reversible. These properties of the ^{35}S-PrP^{res} conversion product are indistinguishable from those of PrP^{Sc} generated in scrapie-infected animals (Kocisko et al. 1996). The fact that the conversion reactions reflect fundamental aspects of TSE biology suggest they are relevant to the molecular pathogenesis of these diseases. In addition, the cell-free conversion reactions have been used for practical purposes such as predicting host susceptibilities to TSE agents (Raymond et al. 1997, 2000; Bossers et al. 2000) and screening for anti-TSE compounds (Caughey et al. 1998b, 2003; Chabry et al. 1998; Demaimay et al. 1998; Maxson et al. 2003).

6
Mechanistic Insights from Cell-Free Reactions

The mechanism by which PrP^{C} is converted into PrP^{Sc} remains unclear. However, a variety of observations warrant consideration in evaluating mechanistic models of PrP^{res} formation. The reaction is dependent upon time and the concentrations of input PrP^{res} and PrP^{sen} (Caughey et al. 1995), although the dependence on these parameters can be complex and condition-dependent (Caughey et al. 1995; Horiuchi and Caughey 1999b). Converting activity is detected in association with PrP^{res} polymers but not with soluble, monomeric forms of PrP (Caughey et al. 1995, 1997; J. Silveira and B. Caughey, unpublished results). The newly converted PrP^{res} remains associated with the PrP^{res} aggregates and is not released as a soluble entity without GdnHCl treatment (Bessen et al. 1997; Callahan et al. 2001). The reaction efficiency is modulated by a variety of factors which can affect protein folding including GdnHCl (Kocisko et al. 1994; Zou and Cashman 2002), urea (DebBurman et al. 1997), certain detergents (Raymond et al. 1997; Zou and Cashman 2002), chaperones (Tatzelt et al. 1996; DebBurman et al. 1997), or, in the absence of denaturants or detergents, sulfated glycans, and elevated temperature (Wong et al. 2001b). Kinetic analyses provided evidence that the conversion process can be separated into two stages, first the binding of PrP^{sen} to PrP^{Sc}, and then a slower conversion of the bound PrP^{sen} to

PrP^{res} (DebBurman et al. 1997; Bessen et al. 1997; Horiuchi and Caughey 1999b). Comparisons of the GdnHCl-containing (Kocisko et al. 1994) vs. GdnHCl and detergent-free reaction conditions (Wong et al. 2001b) suggest that the kinetics of the latter reactions are enhanced both at the level of binding (Maxson et al. 2003) and acquisition of PK resistance (Wong et al. 2001b). Zou and Cashman (2002) have recently found a stimulatory effect by pretreatment of brain PrP^C with acid pH and GdnHCl followed by exposure to a low concentration of SDS which perhaps generates 'recruitable intermediates' of PrP^C for PrP^{res} formation. Cyclic sonication greatly assisted the generation of new PrP^{res} in another cell-free system (Saborio et al. 2001), although other labs have had difficulty reproducing this level of amplification possibly due to technical reasons. Together, these observations, and the formation of amyloid fibril polymers by PrP^{res}, are consistent with an autocatalytic or templated nucleated polymerization mechanism. However, the fact that PrP^{res} usually induces the conversion of only substoichiometric quantities of PrP^{sen} in current cell-free reactions makes the reaction less continuous than typical nucleated polymerizations of proteins or peptides, although more recent renditions of the reaction have made improvements in addressing this problem (Saborio et al. 2001; Zou and Cashman 2002). This may be a technical problem rather than a fundamental limitation of the reaction mechanism. On the other hand, since PrP^{res} forms continuously for long periods of time in vivo, there may be important elements of the mechanism, such as cofactors or micro-environments, that remain to be elucidated (DebBurman et al. 1997; Saborio et al. 1999).

6.1
Role of the Disulfide Bond

A subject of current debate is the role of the PrP disulfide bond that links the second and third helices in the ordered C-terminal domain of PrP^C. Somerville et al. first reported that scrapie infectivity is sensitive to the disulfide reducing agent 2-mercaptoethanol but only in the presence of SDS (Somerville et al. 1980). Biochemical analyses indicated that PrP^{Sc}, like PrP^C, contains an intact disulfide bond (Turk et al. 1988). Furthermore, the cell-free conversion reaction is inhibited by treatment of either PrP^{sen} or PrP^{res} with the reducing agent dithiothreitol (Herrmann and Caughey 1998). These complementary lines of evidence suggest that preservation of the disulfide bond is important in infectivity and PrP^{Sc}

formation. However, recent experiments by Jackson et al. showed that reduction of the disulfide bond aids in the conversion of recombinant PrP^{sen} to a high β-sheet, PK-resistant, fibril-forming state similar to PrP^{Sc} (Jackson et al. 1999). To date, this so-called 'β-PrP' has not been shown to be infectious, but its discovery has rekindled questions as to whether reduction of the disulfide bond (perhaps only transiently in vivo) might play a role in the PrP^{C} to PrP^{res} conversion mechanism. In this light, a new model of PrP^{res} formation was proposed suggesting that infectious prions are composed of PrP^{Sc} polymers linked by intermolecular disulfide bonds and that PrP^{C} to PrP^{Sc} conversion may involve not only a conformational transition but also a thiol–disulfide exchange reaction between the terminal thiolate of such a PrP^{Sc} polymer and the disulfide bond of a PrP^{C} monomer (Welker et al. 2001). However, a recent test of this hypothesis using cell-free reactions provided evidence that conversion of oxidized, disulfide-intact PrP^{sen} to PrP^{res} can occur without the temporary breakage and subsequent reformation of the disulfide bonds (Welker et al. 2002).

6.2 Sites of Interaction Between PrP^{sen} and PrP^{res}

Anti-PrP antibodies have been used to begin mapping the surfaces on PrP^{sen} that are involved in its initial binding to PrP^{Sc} (Horiuchi and Caughey 1999b) (Fig. 2). So far, only antibodies against the C terminus of PrP (α219–232) have inhibited binding. However, removal of the α219–232 epitope from PrP^{sen} did not eliminate binding, suggesting that it is not the epitope itself that is required, but residues close to it in space that are sterically hindered by antibody binding. Surfaces that fit this description include the extended chain of residues ~119–140, the loop of residues ~165–174 between the second β-strand and the second α-helix, and helical residues ~206–223 (Fig. 2) (Horiuchi and Caughey 1999a, 1999b). Antibodies to other regions of the PrP^{C} molecule can block PrP^{Sc} formation in cell culture (Enari et al. 2001; Peretz et al. 2001b) and animals (Heppner et al. 2001; White et al. 2003), although the effect of these antibodies on direct interactions between PrP^{C} and PrP^{Sc} remains to be determined.

Peptide inhibition studies have also pointed to the importance of these regions, as peptides corresponding to residues 119–136 (Chabry et al. 1998, 1999), 166–179 and 200–223 (Horiuchi et al. 2001b) inhibit both

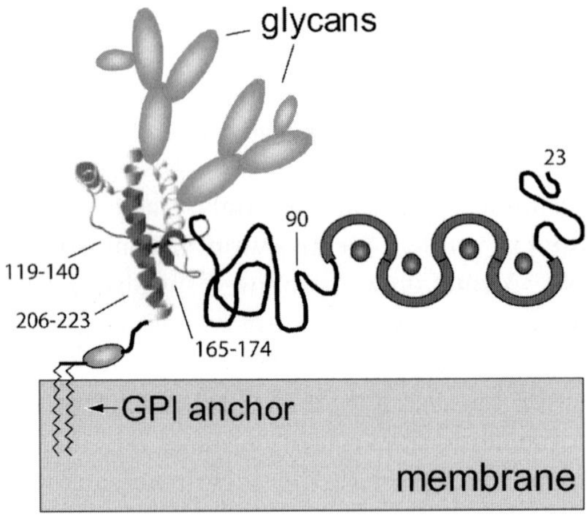

Fig. 2 Potential PrP^{res} binding sites (Horiuchi and Caughey 1999b) on the NMR structure of hamster PrP-sen (Liu et al. 1999). The structure is shown modeled in a GPI-anchored, membrane-associated state. Amino-terminal octapeptide repeat sequences (*shaded semicircles*) are shown with bound copper (*spheres*). Regions that might be blocked by binding of an antibody raised against residues 219–232 that inhibits binding to PrP^{res} are indicated: 119–140, 165–174, and 206–223. Note that these regions are clustered near the C terminus and form a surface on the bottom of the molecule oriented towards the membrane that might be sterically occluded from interacting with large, exogenous aggregates of PrP^{res} . This surface could become accessible upon release of PrP^{sen} from the membrane or insertion of PrP^{res} into a contiguous membrane

the binding and conversion reactions. Peptides corresponding to residues 109–141 and 90–145 can also form protease-resistant complexes with PrP^{C} (Kaneko et al. 1995, 1997b). Amino acid substitutions in these regions have been shown to affect PrP^{res} formation in scrapie-infected cells (Kaneko et al. 1997c). The authors of the study proposed the existence of an unidentified factor, protein X, to explain the inhibition of PrP^{res} formation by point mutations in these regions. However, it also seems possible that the mutations affect the direct binding and/or conversion of PrP^{C} to PrP^{Sc}. The ability of such mutants to inhibit PrP^{Sc} formation in *trans* may be due to an ability to bind wild-type PrP^{res}, and, without converting themselves, to interfere with the binding and/or conversion of wild type PrP^{C} as has been shown for heterologous PrP^{sen} molecules (Priola et al. 1994; Horiuchi et al. 2000).

6.3 Conversions with Membrane-Bound PrP^{sen}

All forms of the cell-free conversion discussed above involve the use of either purified substrates and/or the presence of detergents/denaturants, conditions in which the PrP^{sen} molecules would be extracted from cellular membranes and would therefore potentially be in a non-native state. Conceivably, membrane association of PrP^{sen} could affect its interactions with PrP^{res} in a number of ways such as conformational alterations in PrP^{sen}, as has been shown with recombinant PrP^{sen} (Morillas et al. 1999; Sanghera and Pinheiro 2002), or steric effects. To model the molecular processes that might occur during the initial infection of cells with exogenous TSE agents and to examine the effect of membrane association on the conversion of PrP^{sen} to PrP^{res}, a cell-free conversion reaction was developed that more closely approximates physiological conditions (Baron et al. 2002). These reactions contained purified rafts, or detergent resistant membranes (DRMs), as a source of membrane-bound PrP^{sen} and brain microsomes from scrapie-infected mice as a source of PrP^{res}. Interestingly, DRM-associated PrP^{sen} was not converted to PrP^{res} until the PrP^{sen} was released from DRMs by treatment with phosphatidylinositol-specific phospholipase C (PI-PLC) or the combined membrane fractions were treated with the membrane-fusing agent polyethylene glycol (PEG) (Fig. 3).

These observations have been recapitulated in a more defined system using purified PrP^{sen} reconstituted into model raft liposomes (Baron and Caughey 2003). Liposome studies have also shown that PrP^{sen} can bind to model membranes in a manner independent of the GPI-anchor (Morillas et al. 1999; Sanghera and Pinheiro 2002; Baron and Caughey 2003). GPI-anchor-deficient PrP^{sen}, in contrast to PrP^{sen} bound via its GPI-anchor, was readily converted by exogenous PrP^{res} molecules without PEG or PI-PLC treatment showing that PrP^{sen} membrane association per se does not prevent conversion (Baron and Caughey 2003). Together, these studies suggest that membrane association directed by the GPI-anchor inhibits conversion of membrane-bound GPI-anchored PrP^{sen} by exogenous PrP^{res} (Fig. 4). It is possible that when tethered to the membrane via the GPI-anchor, the PrP^{res} binding site on PrP^{sen} is occluded and incapable of binding to PrP^{res} in a manner leading to conversion until either the PrP^{res} is inserted into a membrane contiguous with PrP^{sen} or the PrP^{sen} is extracted from the membrane. These results raise the possibility that generation of new PrP^{res} during the spread of TSE infec-

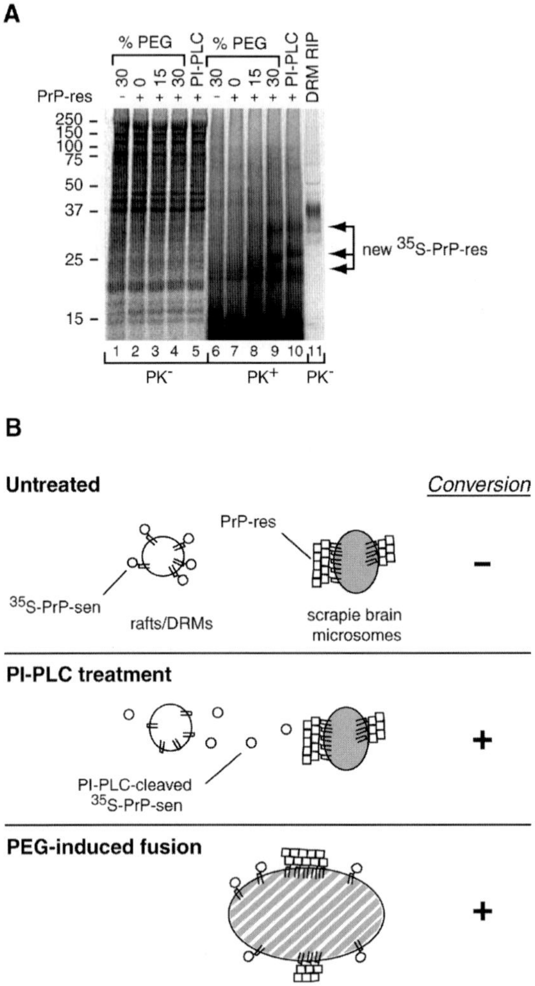

Fig. 3A, B PEG and PI-PLC assist cell-free conversion of DRM-associated PrP^{sen}. **A** PEG- and PI-PLC-assisted [^{35}S]DRM conversion reactions. Samples were treated with various concentrations of PEG or PI-PLC as indicated and cell-free conversion was performed using [^{35}S]DRMs from radiolabeled neuroblastoma cells and crude microsomes from the brains of normal (PrP^{res} negative lanes) or scrapie-infected (PrP^{res} positive lanes) mice. PrP^{sen} was immunoprecipitated from one-fifth equivalents of the [^{35}S]DRMs added to the reaction for comparison (DRM RIP, *lane 11*). *Arrows* indicate PK-resistant [^{35}S]PrP^{res} bands (*lanes 9 and 10*). Adapted from (Baron et al. 2002). **B** Schematic depiction of events occurring during DRM/microsome conversion reactions in **A** for reactions treated with 0% PEG (untreated), PI-PLC (PI-PLC treatment), or 30% PEG (PEG-induced fusion)

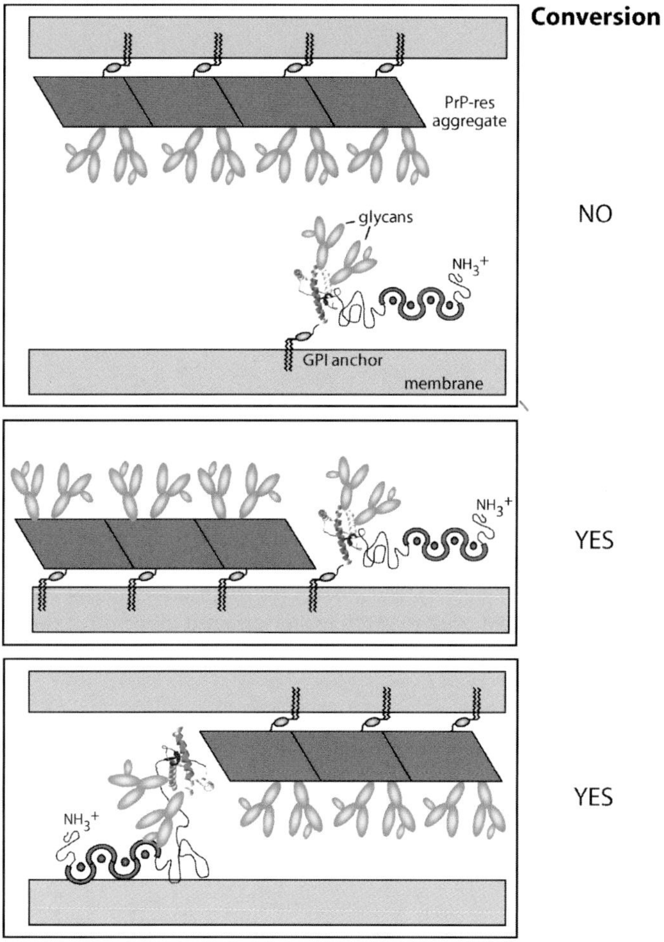

Fig. 4 Summary of cell-free conversion reactions with membrane-associated forms of PrP^{sen}. Potential regions of PrP^{sen} involved in binding PrP^{res} indicated in Fig. 2 are shown in color. In the *top panel*, PrP^{sen} bound to a membrane via the GPI anchor resists conversion by exogenous PrP^{res}. In the *middle panel*, this barrier to conversion is overcome by insertion of PrP^{res} into a membrane contiguous with GPI-anchored PrP^{sen}. In the *bottom panel*, GPI anchor-deficient PrP^{sen} associates with the membrane via amino acid residues in the N terminus, possibly allowing re-orientation of the molecule to allow exogenous PrP^{res} to interact with residues comprising the PrP^{res}-binding site on PrP^{sen} and induce conversion

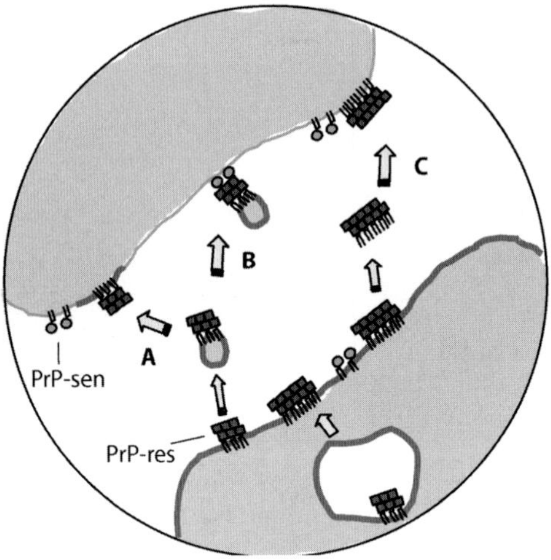

Fig. 5 Possible mechanisms for transmission of PrP^{res} (*squares*) from an infected cell (thick membrane) to an uninfected cell (thin membrane). (*A*) Shedding of PrP^{res}-containing membrane microparticles that fuse with the recipient cell membrane. (*B*) PrP^{res} shed from an infected cell might bind and convert GPI anchor-deficient PrP^{sen} bound to the surface of uninfected cells. (*C*) GPI anchor-directed insertion of extracellular PrP^{res} into recipient cell membrane, also known as 'painting'. PrP^{sen} (*circles*). (Adapted from Baron et al. 2002)

tion between cells requires: (a) removal of PrP^{sen} from target cells; (b) an exchange of membranes between cells (Fig. 5, A); or (c) insertion of incoming PrP^{res} into the raft domains of recipient cells (Fig. 5, C).

6.4
Role of Helix 1 Salt Bridges

The potential role of salt bridges in helix 1 (residues 144–153, hamster) of PrP^{sen} in the conversion process has attracted attention due to modeling studies of Morrissey and Shakhnovich (1999). These authors proposed that helix 1 is stabilized by two aspartic acid-arginine (D144–R147 and D148–R151) intra-helix salt bridge pairs but might favorably adopt conformations necessary to form intermolecular β-sheet aggregates that are stabilized by intermolecular salt bridges between helix 1 aspartic acid and arginine residues of adjacent molecules. The role of these salt bridges in hamster PrP^{sen} stabilization and conversion to PrP^{res} was recently investigated by mutation of the two aspartic acid residues individually and together (Speare et al. 2003). Spectroscopic analyses of the mutant proteins indicated that the salt bridges do not substantially stabilize PrP^{sen}. However, cell-free conversions with the mutant proteins under conditions favorable to salt bridge formation were up to fourfold more efficient than the wild type protein, suggesting that the helix 1 salt bridges stabilize PrP^{sen} from converting to PrP^{res}.

7
Species Barriers and Interactions Between Heterologous PrP Molecules

TSE diseases are most easily transmitted among members of the same or closely related species. Transmission to more distantly related species is also possible, but normally involves a progressive adaptation during serial passage in the new host, where the incubation period of the disease is longer in the first few passages, but eventually stabilizes to a decreased, predictable value. This difficultly associated with transmitting the disease between species has been termed the 'species barrier' (Dickinson 1976). In addition to scrapie from sheep and goats, BSE from cattle has been transferred to mice to produce TSE disease (Fraser et al. 1992). A variety of sources have been used to create a collection of mouse and hamster-adapted TSE strains, enabling the study of pathogenesis and characterization of the infectious agent in these rodent models. In light of recent concerns regarding interspecies transmission of the BSE and CWD agents, determination of the factors controlling the susceptibility of hosts to TSE agents of other species is of key importance. Experiments using transgenic mice and cell culture models of TSE infection, as well as cell-free studies, have illustrated a requirement for compatibility at the primary structure level between the TSE agent-associated PrP^{res} and host PrP^{sen} to allow initiation of infection and propagation of PrP^{res} (Kocisko et al. 1995; Raymond et al. 1997; for a review see Priola 2001; Asante and Collinge 2001). It is notable that these amino acid sequence requirements can be specific to each PrP^{res}/PrP^{sen} combination (Priola et al. 2001), confounding predictions of interspecies transmissions based on PrP sequence comparisons alone. Interestingly, the presence of heterologous PrP^{sen} molecules can have a significant inhibitory effect on these processes in a phenomenon referred to as 'interference' (Prusiner et al. 1990; Priola et al. 1994; Horiuchi et al. 2000).

7.1 Relationship of Efficiencies of Conversion to Interspecies Transmission

Cell-free binding and conversion reactions have been used to directly test the molecular compatibility between PrP^{sen} derived from one species and PrP^{res} derived from another. Since TSE infection can be acquired by different routes (e.g., ingestion, direct intracerebral inoculation), factors other than PrP^{sen}/PrP^{res} sequence compatibility may influence the outcome following TSE agent exposure. Hence, these cell-free analyses probably most closely correlate with relative intracerebral transmission titers. Interspecies conversion reactions have been performed with a number of PrP^{sen}/PrP^{res} combinations, and profound sequence specificity is often observed (Kocisko et al. 1995; Raymond et al. 1997, 2000; Bossers et al. 1997, 2000; Chabry et al. 1999). When assayed in parallel using the same PrP^{res} preparation, several fold (5- to >50-fold) stronger conversion efficiencies have been observed with PrP^{sen} molecules from highly susceptible animals versus clearly resistant species/genotypes. Intermediate efficiencies (two- to fourfold weaker than homologous) have been observed with PrP^{sen} from animals that are known to be susceptible, but apparently less so than the original host species. Based on the available information, it seems that the log of the relative intracerebral transmission titer might be roughly proportional to the relative cell-free conversion efficiency on a linear scale (Raymond et al. 2000). However, much more quantitative transmission data between various species will be required to establish the fit between these parameters. Nonetheless, the requirement for molecular compatibility between different PrP^{res} and PrP^{sen} sequences, as reflected both in vivo and in vitro, appears to be an important factor in the transmission process.

7.2 BSE and CWD

The cell-free conversion assay has been used to assess the likely susceptibility of various hosts to TSE agents from different source species or genotypes (Raymond et al. 1997, 2000; Bossers et al. 2000), especially under circumstances where direct transmission studies cannot be performed due to ethical considerations (e.g., human transmissions), experimental restrictions (e.g., lack of a transgenic model for CWD), or time required to complete such studies. For instance, little is known about the transmissibility of CWD of deer and elk to other noncervid species except for ferrets (Bartz et al. 1998; Miller et al. 2000). In cell-free conversion reactions, we recently found that the CWD-associated PrP^{res} (PrP^{CWD}) of cervids readily induces the conversion of cervid PrP^{sen} molecules to the protease-resistant state in accordance with the known transmissibility of CWD between cervids (Fig. 6) (Raymond et al. 2000). Both allelic variants of elk PrP^{sen} (Met or Leu at codon 132) were converted with similar efficiencies by PrP^{CWD} from elk, mule deer, or white-tail deer, suggesting that the over-representation of CWD-infected elk homozygous for PrP codon Met132 (O'Rourke et al. 1999) is not explained by direct incompatibility of the Leu132 PrP^{sen} with the PrP^{CWD} used in these experiments. However, this over-representation might be explained by a predominance of the Met132 codon in the animals sampled (O'Rourke et al. 1999). In contrast, PrP^{CWD}-induced conversions of human and bovine PrP^{sen} were much less efficient, and conversion of ovine PrP^{sen} was intermediate. These results demonstrate a barrier at the molecular level that should limit, but not necessarily eliminate, the susceptibility of these noncervid species to CWD. These data also seem to correlate with preliminary observations to date on the experimental transmission of mule deer-derived CWD to cattle whereby only three of 13 calves inoculated intracerebrally succumbed to TSE infection at 24–27 months post-inoculation (Hamir et al. 2001). At the time of publication 3 years after challenge, the remaining 10 calves were alive and healthy, consistent with the predictions from cell-free conversion studies of a limited susceptibility of cattle to the CWD agent that may be overcome with low frequency by a high dose infection via the most efficient route (Raymond et al. 2000). It is important to note that the cell-free reactions above do not predict potential changes in the host range of TSE agents after passage through 'intermediate' species, a phenomenon that has been shown to occur with CWD (Bartz et al. 1998).

The origins of the CWD and BSE agents remain uncertain but have been proposed in both cases to be derived by adaptation of some form of sheep scrapie (Williams and Miller 2002; Baylis et al. 2002). Hence, it is also of interest to measure the compatibility of the cervid and bovine PrP^{sen} molecules with PrP^{res} prepared from other species, especially sheep. Of the ovine and bovine PrP^{res} types, ovine PrP^{res} (from sheep of the ARQ genotype) induced the strongest conversion of cervid PrP^{sen} molecules, but even these conversions were at least fivefold less efficient than the homologous conversion of ovine ARQ-PrP^{sen} (Raymond et al.

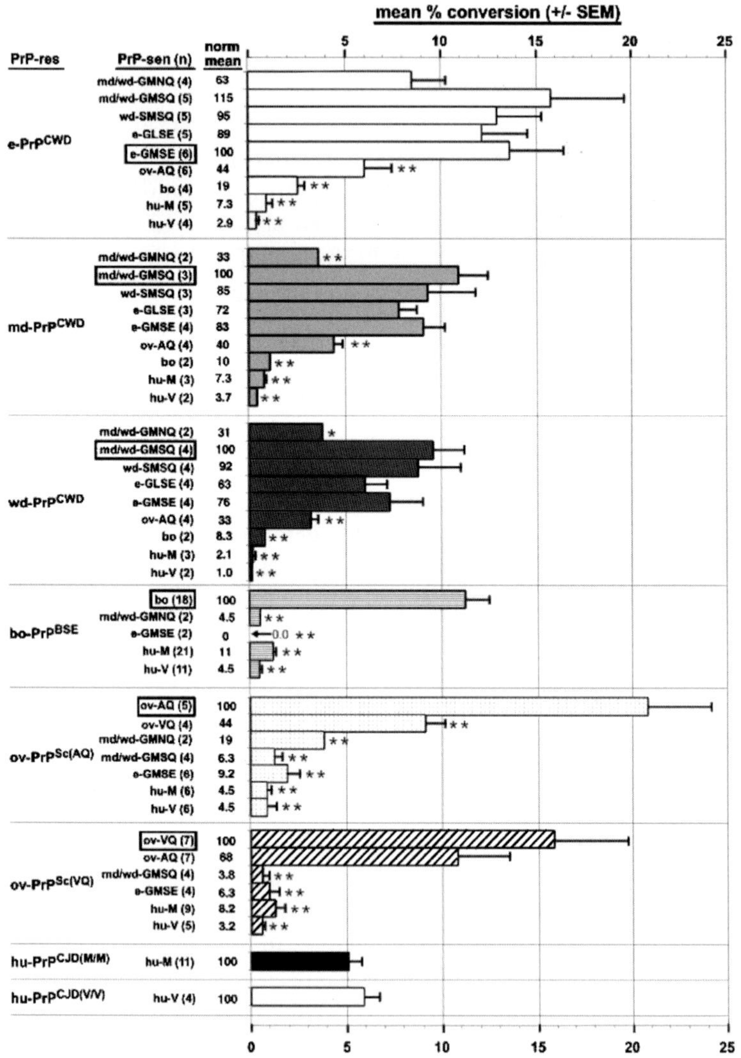

Fig. 6 Efficiencies of conversion reactions between homologous and heterologous isoforms of PrP. The conversion efficiencies of $[^{35}S]$-PrP^{sen} proteins in cell-free reactions with equivalent amounts of different PrP^{res} molecules are shown. The results show the mean percentage of the input 24–27 kDa $[^{35}S]$-PrP^{sen} converted to the 17–20 kDa PK-resistant $[^{35}S]$-PrP bands. PrP^{CJD}-M/M and PrP^{CJD}-V/V designate human PrP^{res} derived from the brains of CJD patients homozygous for methionine or valine at residue 129, respectively. Within the groups of values for each type of PrP^{res}, the statistical significance of the difference of the means relative to the mean percentage conversion of PrP^{sen} of the same species (*boxed*) was assessed with a one-way ANOVA with Dunnett's multiple comparison test: **P*<0.05; ***P*<0.01. The 'norm mean' column shows means within each PrP^{res} group normalized to the homologous conversion (*boxed*). (Adapted from Raymond et al. 2000)

2000), implying the existence of a finite molecular barrier to efficient transmission of sheep scrapie into cervid species. Conversely, ovine PrP^{res} (from sheep of the VRQ genotype) induced the conversion of bovine PrP^{sen} with comparable efficiency to ovine PrP^{sen} of susceptible genotypes (ARQ or VRQ) (Raymond et al. 1997), again consistent with the known susceptibility of cattle to sheep scrapie (Gibbs et al. 1990; Cutlip et al. 1994).

To gain insight into the potential transmissibility of BSE and scrapie to humans, conversions of human PrP^{sen} molecules of either the methionine or valine allelic form at codon 129 (corresponds to the polymorphic codon 132 of elk) have also been performed with PrP^{BSE} and ovine PrP^{Sc} from sheep of two susceptible genotypes (ARQ and VRQ). These conversions of human PrP^{sen} by bovine and ovine PrP^{res} are all more than 10-fold less efficient compared to the corresponding homologous conversions and are close to the limit of detection for conversion product (Raymond et al. 1997, 2000). Thus, the inherent ability of the BSE and scrapie agents to affect humans following equivalent exposure may be finite but low. However, it does appear that the BSE agent has caused vCJD in a relatively small number of people. Interestingly, all vCJD patients at the time of writing are Met129 homozygous, and similarly PrP^{BSE} was found to convert Met129 human PrP^{sen} more efficiently than the Val129 variant (Raymond et al. 1997). Given the presumably large number of people exposed to BSE infectivity, the susceptibility of humans may still be very low compared with cattle, which would be consistent with the relatively inefficient conversion of human PrP^{sen} by PrP^{BSE} (Raymond et al. 1997, 2000). Nevertheless, it would seem prudent to take reasonable measures to limit exposure of all species to TSE agents.

7.3 Binding vs. Conversion of Heterologous PrP Molecules

The conversion reaction has been dissected kinetically into two distinct steps: initial binding and conversion to PrP^{res} (as measured by acquisition of protease resistance). The question then arises as to which step is most important in controlling the sequence specificity of the conversion reaction (DebBurman et al. 1997; Horiuchi and Caughey 1999b). Recent experiments using mouse and hamster PrP isoforms indicated that the binding of PrP^{sen} to heterologous PrP^{Sc} can occur much more efficiently than the subsequent conversion to PrP^{res} (Horiuchi et al. 2000). This suggests that the species-specificity of PrP^{res} formation is determined more by the conversion step than the initial binding step. The binding of heterologous, nonconverting PrP^{sen} molecules to PrP^{Sc} does not block the binding of homologous PrP^{sen} , but does interfere with its conversion. This interference effect could explain reductions in the rates of PrP^{res} formation and/or TSE pathogenesis in hosts that co-express heterologous PrP^{C} molecules (Prusiner et al. 1990; Palmer et al. 1991; Priola et al. 1994).

As noted above, there are two current hypotheses regarding the sites of binding and conversion. In one hypothesis, the binding and conversion steps occur at a single site. Alternatively, there may be two types of PrP^{sen} binding sites on PrP^{res}, one that is 'conversion-competent' (say, at the ends of growing fibrils) and another that is not (e.g., along the sides of fibrils) (Caughey et al. 1995, 1997). The initial binding of PrP^{sen} to PrP^{res} at either type of site may occur in a manner that is less amino acid sequence-specific than the further steps required at conversion-competent sites for the conversion of PrP^{sen} to the PK-resistant state (Horiuchi et al. 2000). Studies using mouse/hamster chimeric PrP have shown that the central part of the PrP^{sen} molecule, including three amino acid substitutions at mouse/hamster residues 138/139, 154/155, and 169/170, is important in the conversion of PrP^{sen} to PrP^{res} (Scott et al. 1992; Priola et al. 1994; Kocisko et al. 1995; Priola and Chesebro 1995). However, the precise residues involved may be specific for different PrP^{sen}/PrP^{res} combinations. This has been observed in conversions between hamster and mouse PrP molecules, where although generation of mouse PrP^{res} was strongly influenced by homology between PrP^{sen} and PrP^{res} at amino acid residue 138, homology between PrP^{sen} and PrP^{res} at amino acid residue 155 determined the efficiency of formation of a protease-resistant product induced by hamster PrP^{res} (Priola et al. 2001). Thus, it is possible that critical interactions in the vicinity of these residues on PrP^{sen} and/or PrP^{Sc} occur as part of the conversion step, at least in rodent models of scrapie.

8
TSE Strains

Multiple TSE strains can be distinguished within a single host species or PrP genotype by reproducible differences in incubation period, clinical signs, vacuolar brain pathology and PrP^{res} deposition (for review see

Bruce and Fraser 1991). The existence of TSE strains independent of the host PrP genotype presents a major challenge to the protein-only hypothesis for the TSE agent. However, there is evidence to support the hypothesis that the mechanism for TSE strain propagation is determined by the self-propagation of PrP^{res} molecules that differ in conformation, GdnHCl unfolding, glycoform ratios, polymeric states, and/or ligand associations (Bessen and Marsh 1994; Bessen et al. 1995; Collinge et al. 1996; Telling et al. 1996; Somerville et al. 1997; Safar et al. 1998; Caughey et al. 1998a; Rubenstein et al. 1998; Peretz et al. 2001a, 2002).

The possibility of multiple conformations of PrP^{res} was suggested by observations that PrP^{res} associated with different strains of TSEs are cleaved at different N-terminal sites by PK. This was first documented with the Hyper (HY) and Drowsy (DY) strains of transmissible mink encephalopathy (TME) passaged in hamsters (Bessen and Marsh 1992, 1994; Bartz et al. 2000). After treatment with PK, the PrP^{res} from HY-infected hamsters had a ~2 kDa larger molecular mass than PrP^{res} from DY-infected hamsters in SDS–PAGE analysis. N-terminal sequencing revealed that PK cleaved additional residues from the N terminus of DY PrP^{res} compared to HY PrP^{res}. Since the DY and HY PrP^{res} molecules are derived from the same Syrian hamster PrP^{sen} precursor, it was concluded that the observed difference in cleavage by PK was due to strain-dependent differences in conformation and/or ligand binding.

Fourier transform infrared spectroscopic analyses have further shown that HY and DY PrP^{res} differ dramatically in the β-sheet region (Caughey et al. 1998a). More recently, related, but subtle differences have been observed amongst spectra of various strains of mouse PrP^{Sc} (B. Caughey and G. Raymond, unpublished results). Wadsworth and colleagues have shown that metal ion binding can influence the site of PK cleavage in PrP^{res} (Wadsworth et al. 1999). Evidence for multiple strain differences in PrP^{Sc} conformation has also been obtained with multispectral ultraviolet fluorescence spectroscopy (Rubenstein et al. 1998), denaturant-induced PK-susceptibility (Peretz et al. 2001a), and by comparing antibody epitope exposure caused by denaturant treatments, a technique called conformation-dependent immunoassay (CDI) (Safar et al. 1998, 2002).

Direct evidence that the HY- and DY-associated conformations of hamster PrP^{res} could propagate themselves from the same hamster PrP^{sen} precursor was obtained using a cell-free conversion reaction (Bessen et al. 1995). In these studies, HY and DY PrP^{res} were each incubated with hamster PrP^{sen} and the PrP^{sen} was converted to PK-resistant PrP products with the same ~1–2 kDa difference in molecular mass that distinguishes the PK-treated HY and DY PrP^{res} molecules. This finding suggested that the strain-specific forms of PrP^{res} are faithfully propagated through direct PrP–sen–PrP–res interactions both in vitro and in vivo as had been proposed earlier (Bolton and Bendheim 1988; Prusiner 1991).

Strain-dependent differences in either glycoform ratios (i.e., PrP molecules with none, one or two N-linked glycans) and/or PK cleavage sites in PrP^{res} have been observed with other TSEs, and these differences now serve as a biochemical aid in TSE strain typing (Monari et al. 1994; Collinge et al. 1996; Telling et al. 1996; Somerville et al. 1997; Parchi et al. 1997; Bartz et al. 2000; Asante and Collinge 2001). Strain-dependent variation in the relative proportions of PrP^{res} glycoforms of various murine scrapie strains was first reported by Kascsak and colleagues (Kascsak et al. 1985, 1991). Indeed, this type of analysis has provided corroborative evidence connecting the BSE strain of TSE in cattle to vCJD in humans (Hill et al. 1997; Somerville et al. 1997). Using a similar type of analysis, the transmission of different strains of human CJD into mice leads to the formation of mouse PrP^{res} with different PK cleavage sites (Telling et al. 1996). These observations show that the strain-specific conformations can be propagated between species as well as within species.

A very recent analysis of the PrP^{res} glycoform patterns present in infected cervids found that the patterns for elk were more tightly grouped than those for deer, leading the authors to suggest that the elk and deer examined in this study may have been infected by different strains of CWD agent, perhaps multiple strains in the case of deer (Race et al. 2002). Race et al. also observed a general similarity between the PrP^{res} glycoform patterns found in scrapie-infected sheep and CWD-infected cervids, raising the possibility that CWD arose from sheep scrapie (Race et al. 2002). Consistent with the former proposal, Safar and coworkers found evidence for conformational differences between PrP^{CWD} molecules derived from deer when compared to elk, but not between those derived from mule deer vs. white tail deer (Safar et al. 2002).

9 Conclusions

Despite recent progress in several issues associated with TSE diseases, an understanding of many fundamental aspects is still lacking. Debate persists over the cellular function of PrP^C. The raft association of GPI-anchored PrP^C is consistent with its potential function as a signaling molecule. High-resolution structural data of PrP^{res} aggregates remains scarce, although a new approach through electron crystallography may help to solve this problem. Cell-free studies have continued to provide insights into mechanisms of PrP^{res} formation. Sites of specific interaction between normal and abnormal PrP molecules are being mapped, information that is of obvious value in the design of TSE therapeutics. New cell-free reaction conditions using membrane-bound substrates have been developed that more closely approximate the circumstances under which PrP^C and PrP^{res} interact in vivo. Such reactions have provided clues to factors that might govern the exchange of PrP^{res} between cells and initiation of infection in cells. Furthermore, cell-free conversion reactions have been useful in assessing the molecular compatibilities of PrP^C and PrP^{res} molecules of different species to evaluate the potential risks of interspecies TSE transmissions, a subject of great concern with the recent increases in the incidence of BSE and CWD. However, the precise nature of the TSE infectious agent has still eluded identification. Considerable circumstantial evidence suggests that an abnormal form of PrP is involved. Nevertheless, it is still uncertain which form(s) of abnormal PrP, if any, is/are responsible for the infectious and neurodegenerative characteristics of TSEs. In this regard, highly efficient new methods for the cell-free generation of PrP^{res} might provide answers to some of these crucial questions, although as yet none of these reactions has been shown to produce new TSE infectivity.

Acknowledgements. G.S.B. was supported in part by post-doctoral fellowships from the Natural Sciences and Engineering Research Council of Canada and the Canadian Institutes of Health Research.

References

DEFRA BSE information General statistics-GB HM Government UK Online-Department for Environment Food and Rural Affairs. http://www.defra.gov.uk/animalh/bse/bse-statistics/bse/general.html

UK Monthly CJD statistics UK-Department of Health. http://www.doh.gov.uk/cjd/cjd_stat.htm

Andreoletti O, Lacroux C, Chabert A, Monnereau L, Tabouret G, Lantier F, Berthon P, Eychenne F, Lafond-Benestad S, Elsen JM, Schelcher F (2002) PrP(Sc) accumulation in placentas of ewes exposed to natural scrapie: influence of foetal PrP genotype and effect on ewe-to-lamb transmission. J Gen Virol 83:2607–2616

Asante EA, Collinge J (2001) Transgenic studies of the influence of the PrP structure on TSE diseases. Adv Protein Chem 57:273–311

Baron GS, Caughey B (2003) Effect of Glycosylphosphatidylinositol Anchor-dependent and -independent Prion Protein Association with Model Raft Membranes on Conversion to the Protease-resistant Isoform. J Biol Chem 278:14883–14892

Baron GS, Wehrly K, Dorward DW, Chesebro B, Caughey B (2002) Conversion of raft associated prion protein to the protease-resistant state requires insertion of PrP-res (PrP(Sc)) into contiguous membranes. EMBO J 21:1031–1040

Bartz JC, Bessen RA, McKenzie D, Marsh RF, Aiken JM (2000) Adaptation and selection of prion protein strain conformations following interspecies transmission of transmissible mink encephalopathy. J Virol 74:5542–5547

Bartz JC, Marsh RF, McKenzie DI, Aiken JM (1998) The host range of chronic wasting disease is altered on passage in ferrets. Virology 251:297–301

Baskakov IV, Legname G, Baldwin MA, Prusiner SB, Cohen FE (2002) Pathway complexity of prion protein assembly into amyloid. J Biol Chem 277:21140–21148

Basler K, Oesch B, Scott M, Westaway D, Walchli M, Groth DF, McKinley MP, Prusiner SB, Weissmann C (1986) Scrapie and cellular PrP isoforms are encoded by the same chromosomal gene. Cell 46:417–428

Baylis M, Houston F, Kao RR, McLean AR, Hunter N, Gravenor MB (2002) BSE—a wolf in sheep's clothing? Trends Microbiol 10:563–570

Belay ED, Gambetti P, Schonberger LB, Parchi P, Lyon DR, Capellari S, McQuiston JH, Bradley K, Dowdle G, Crutcher JM, Nichols CR (2001) Creutzfeldt-Jakob disease in unusually young patients who consumed venison. Arch Neurol 58:1673–1678

Bessen RA, Kocisko DA, Raymond GJ, Nandan S, Lansbury PT, Caughey B (1995) Non-genetic propagation of strain-specific properties of scrapie prion protein. Nature 375:698–700

Bessen RA, Marsh RF (1992) Biochemical and physical properties of the prion protein from two strains of the transmissible mink encephalopathy agent. J Virol 66:2096–2101

Bessen RA, Marsh RF (1994) Distinct PrP properties suggest the molecular basis of strain variation in transmissible mink encephalopathy. J Virol 68:7859–7868

Bessen RA, Raymond GJ, Caughey B (1997) In situ formation of protease-resistant prion protein in transmissible spongiform encephalopathy-infected brain slices. J Biol Chem 272:15227–15231

Bolton DC, Bendheim PE (1988) A modified host protein model of scrapie. In: Bock G, Marsh J (eds) Novel Infectious Agents and the Central Nervous System. John Wiley & Sons, Chichester, pp 164–181

Bolton DC, Bendheim PE, Marmorstein AD, Potempska A (1987) Isolation and structural studies of the intact scrapie agent protein. Arch Biochem Biophys 258:579–590

Borchelt DR, Scott M, Taraboulos A, Stahl N, Prusiner SB (1990) Scrapie and cellular prion proteins differ in their kinetics of synthesis and topology in cultured cells. J Cell Biol 110:743–752
Borchelt DR, Taraboulos A, Prusiner SB (1992) Evidence for synthesis of scrapie prion proteins in the endocytic pathway. J Biol Chem 267:16188–16199
Bosque PJ, Ryou C, Telling G, Peretz D, Legname G, DeArmond SJ, Prusiner SB (2002) Prions in skeletal muscle. Proc Natl Acad Sci USA 99:3812–3817
Bossers A, Belt PBGM, Raymond GJ, Caughey B, de Vries R, Smits MA (1997) Scrapie susceptibility-linked polymorphisms modulate the in vitro conversion of sheep prion protein to protease-resistant forms. Proc Natl Acad Sci USA 94:4931–4936
Bossers A, de Vries R, Smits MA (2000) Susceptibility of sheep for scrapie as assessed by in vitro conversion of nine naturally occurring variants of PrP. J Virol 74:1407–1414
Bounhar Y, Zhang Y, Goodyer CG, LeBlanc A (2001) Prion protein protects human neurons against Bax-mediated apoptosis. J Biol Chem 276:39145–39149
Brown DR, Qin K, Herms JW, Madlung A, Manson J, Strome R, Fraser PE, Kruck T, von Bohlen A, Schulz-Schaeffer W, Giese A, Westaway D, Kretzschmar H (1997) The cellular prion protein binds copper in vivo. Nature 390:684–687
Brown DR, Wong BS, Hafiz F, Clive C, Haswell SJ, Jones IM (1999) Normal prion protein has an activity like that of superoxide dismutase. Biochem J 344 Pt 1:1–5
Bruce ME, Fraser H (1991) Scrapie strain variation and its implications. Curr Top Microbiol Immunol 172:125–138
Bruce ME, Will RG, Ironside JW, McConnell I, Drummond D, Suttie A, McCardle L, Chree A, Hope J, Birkett C, Cousens S, Fraser H, Bostock CJ (1997) Transmissions to mice indicate that 'new variant' CJD is caused by the BSE agent. Nature 389:498–501
Bueler H, Aguzzi A, Sailer A, Greiner RA, Autenried P, Aguet M, Weissmann C (1993) Mice devoid of PrP are resistant to scrapie. Cell 73:1339–1347
Callahan MA, Xiong L, Caughey B (2001) Reversibility of scrapie-associated prion protein aggregation. J Biol Chem 276:28022–28028
Cashman NR, Loertscher R, Nalbantoglu J, Shaw I, Kascsak RJ, Bolton DC, Bendheim PE (1990) Cellular isoform of the scrapie agent protein participates in lymphocyte activation. Cell 61:185–192
Caughey B (2001) Interactions between prion protein isoforms: the kiss of death? Trends Biochem Sci 26:235–242
Caughey B, Kocisko DA, Raymond GJ, Lansbury PT, Jr. (1995) Aggregates of scrapie-associated prion protein induce the cell-free conversion of protease-sensitive prion protein to the protease-resistant state. Chem Biol 2:807–817
Caughey B, Lansbury PT Jr (2003) Protofibrils, Pores, Fibrils, and Neurodegeneration: Separating the Responsible Protein Aggregates from the Innocent Bystanders. Annu Rev Neurosci 26:267–298
Caughey B, Neary K, Buller R, Ernst D, Perry LL, Chesebro B, Race RE (1990) Normal and scrapie-associated forms of prion protein differ in their sensitivities to phospholipase and proteases in intact neuroblastoma cells. J Virol 64:1093–1101

Caughey B, Race RE, Ernst D, Buchmeier MJ, Chesebro B (1989) Prion protein biosynthesis in scrapie-infected and uninfected neuroblastoma cells. J Virol 63:175–181

Caughey B, Race RE, Vogel M, Buchmeier MJ, Chesebro B (1988) In vitro expression in eukaryotic cells of a prion protein gene cloned from scrapie-infected mouse brain. Proc Natl Acad Sci USA 85:4657–4661

Caughey B, Raymond GJ (1991) The scrapie-associated form of PrP is made from a cell surface precursor that is both protease- and phospholipase-sensitive. J Biol Chem 266:18217–18223

Caughey B, Raymond GJ, Bessen RA (1998a) Strain-dependent differences in beta-sheet conformations of abnormal prion protein. J Biol Chem 273:32230–32235

Caughey B, Raymond GJ, Ernst D, Race RE (1991a) N-terminal truncation of the scrapie-associated form of PrP by lysosomal protease(s): implications regarding the site of conversion of PrP to the protease-resistant state. J Virol 65:6597–6603

Caughey B, Raymond GJ, Kocisko DA, Lansbury PT, Jr. (1997) Scrapie infectivity correlates with converting activity, protease resistance, and aggregation of scrapie-associated prion protein in guanidine denaturation studies. J Virol 71:4107–4110

Caughey B, Raymond LD, Raymond GJ, Maxson L, Silveira J, Baron GS (2003) Inhibition of protease-resistant prion protein accumulation in vitro by curcumin. J Virol 77:5499–5502

Caughey BW, Dong A, Bhat KS, Ernst D, Hayes SF, Caughey WS (1991b) Secondary structure analysis of the scrapie-associated protein PrP 27–30 in water by infrared spectroscopy. Biochemistry 30:7672–7680

Caughey WS, Raymond LD, Horiuchi M, Caughey B (1998b) Inhibition of protease-resistant prion protein formation by porphyrins and phthalocyanines. Proc Natl Acad Sci USA 95:12117–12122

Chabry J, Caughey B, Chesebro B (1998) Specific inhibition of in vitro formation of protease-resistant prion protein by synthetic peptides. J Biol Chem 273:13203–13207

Chabry J, Priola SA, Wehrly K, Nishio J, Hope J, Chesebro B (1999) Species-independent inhibition of abnormal prion protein (PrP) formation by a peptide containing a conserved PrP sequence. J Virol 73:6245–6250

Chernoff YO, Lindquist SL, Ono B, Inge-Vechtomov SG, Liebman SW (1995) Role of the chaperone protein Hsp104 in propagation of the yeast prion-like factor [psi+]. Science 268:880–884

Chesebro B (1999) Prion protein and the transmissible spongiform encephalopathy diseases. Neuron 24:503–506

Chiarini LB, Freitas AR, Zanata SM, Brentani RR, Martins VR, Linden R (2002) Cellular prion protein transduces neuroprotective signals. EMBO J 21:3317–3326

Chiesa R, Drisaldi B, Quaglio E, Migheli A, Piccardo P, Ghetti B, Harris DA (2000) Accumulation of protease-resistant prion protein (PrP) and apoptosis of cerebellar granule cells in transgenic mice expressing a PrP insertional mutation. Proc Natl Acad Sci USA 97:5574–5579

Chiesa R, Piccardo P, Ghetti B, Harris DA (1998) Neurological illness in transgenic mice expressing a prion protein with an insertional mutation. Neuron 21:1339–1351

Collinge J, Palmer MS, Dryden AJ (1991) Genetic predisposition to iatrogenic Creutzfeldt–Jakob disease. Lancet 337:1441–1442

Collinge J, Sidle KC, Meads J, Ironside J, Hill AF (1996) Molecular analysis of prion strain variation and the aetiology of 'new variant' CJD. Nature 383:685–690

Comincini S, Foti MG, Tranulis MA, Hills D, Di Guardo G, Vaccari G, Williams JL, Harbitz I, Ferretti L (2001) Genomic organization, comparative analysis, and genetic polymorphisms of the bovine and ovine prion Doppel genes (PRND). Mamm Genome 12:729–733

Cutlip RC, Miller JM, Race RE, Jenny AL, Katz JB, Lehmkuhl HD, DeBey BM, Robinson MM (1994) Intracerebral transmission of scrapie to cattle. J Infect Dis 169:814–820

DebBurman SK, Raymond GJ, Caughey B, Lindquist S (1997) Chaperone-supervised conversion of prion protein to its protease-resistant form. Proc Natl Acad Sci USA 94:13938–13943

Demaimay R, Harper J, Gordon H, Weaver D, Chesebro B, Caughey B (1998) Structural aspects of Congo red as an inhibitor of protease-resistant prion protein formation. J Neurochem 71:2534–2541

Dickinson AG (1976) Scrapie in sheep and goats. In: Kimberlin RH (eds) Slow virus diseases of animals and man. North-Holland Publishing Company, Amsterdam, pp 209–241

Dickinson AG, Meikle VM, Fraser H (1968) Identification of a gene which controls the incubation period of some strains of scrapie agent in mice. J Comp Pathol 78:293–299

Diringer H, Gelderblom H, Hilmert H, Ozel M, Edelbluth C, Kimberlin RH (1983) Scrapie infectivity, fibrils and low molecular weight protein. Nature 306:476–478

Donne DG, Viles JH, Groth D, Mehlhorn I, James TL, Cohen FE, Prusiner SB, Wright PE, Dyson HJ (1997) Structure of the recombinant full-length hamster prion protein PrP(29–231): the N terminus is highly flexible. Proc Natl Acad Sci USA 94:13452–13457

Eigen M (1996) Prionics or the kinetic basis of prion diseases. Biophys Chem 63:A1–18

Enari M, Flechsig E, Weissmann C (2001) Scrapie prion protein accumulation by scrapie-infected neuroblastoma cells abrogated by exposure to a prion protein antibody. Proc Natl Acad Sci USA 98:9295–9299

Flechsig E, Shmerling D, Hegyi I, Raeber AJ, Fischer M, Cozzio A, von Mering C, Aguzzi A, Weissmann C (2000) Prion protein devoid of the octapeptide repeat region restores susceptibility to scrapie in PrP knockout mice. Neuron 27:399–408

Fraser H, Bruce ME, Chree A, McConnell I, Wells GA (1992) Transmission of bovine spongiform encephalopathy and scrapie to mice. J Gen Virol 73 (Pt 8):1891–1897

Gajdusek DC (1988) Transmissible and non-transmissible amyloidoses: autocatalytic post-translational conversion of host precursor proteins to beta-pleated sheet configurations. J Neuroimmunol 20:95–110

Giaccone G, Verga L, Bugiani O, Frangione B, Serban D, Prusiner SB, Farlow MR, Ghetti B, Tagliavini F (1992) Prion protein preamyloid and amyloid deposits in Gerstmann–Straussler–Scheinker disease, Indiana kindred. Proc Natl Acad Sci USA 89:9349–9353

Gibbs CJ, Jr., Safar J, Ceroni M, Di Martino A, Clark WW, Hourrigan JL (1990) Experimental transmission of scrapie to cattle. Lancet 335:1275

Graner E, Mercadante AF, Zanata SM, Forlenza OV, Cabral AL, Veiga SS, Juliano MA, Roesler R, Walz R, Minetti A, Izquierdo I, Martins VR, Brentani RR (2000) Cellular prion protein binds laminin and mediates neuritogenesis. Brain Res Mol Brain Res 76:85–92

Griffith JS (1967) Self-replication and scrapie. Nature 215:1043–1044

Guiroy DC, Liberski PP, Williams ES, Gajdusek DC (1994) Electron microscopic findings in brain of Rocky Mountain elk with chronic wasting disease. Folia Neuropathol 32:171–173

Guiroy DC, Williams ES, Song KJ, Yanagihara R, Gajdusek DC (1993) Fibrils in brain of Rocky Mountain elk with chronic wasting disease contain scrapie amyloid. Acta Neuropathol (Berl) 86:77–80

Guiroy DC, Williams ES, Yanagihara R, Gajdusek DC (1991) Topographic distribution of scrapie amyloid-immunoreactive plaques in chronic wasting disease in captive mule deer (Odocoileus hemionus hemionus). Acta Neuropathol (Berl) 81:475–478

Hamir AN, Cutlip RC, Miller JM, Williams ES, Stack MJ, Miller MW, O'Rourke KI, Chaplin MJ (2001) Preliminary findings on the experimental transmission of chronic wasting disease agent of mule deer to cattle. J Vet Diagn Invest 13:91–96

Hegde RS, Tremblay P, Groth D, DeArmond SJ, Prusiner SB, Lingappa VR (1999) Transmissible and genetic prion diseases share a common pathway of neurodegeneration. Nature 402:822–826

Heppner F, Musahl C, Arrighi I, Klein MA, Rulicke T, Oesch B, Zinkernagel RM, Kalinke U, Aguzzi A (2001) Prevention of scrapie pathogenesis by transgenic expression of anti-prion protein antibodies. Science 294:178–189

Herrmann LM, Caughey B (1998) The importance of the disulfide bond in prion protein conversion. Neuroreport 9:2457–2461

Hill AF, Antoniou M, Collinge J (1999) Protease-resistant prion protein produced in vitro lacks detectable infectivity. J Gen Virol 80 (Pt 1):11–14

Hill AF, Desbruslais M, Joiner S, Sidle KC, Gowland I, Collinge J, Doey LJ, Lantos P (1997) The same prion strain causes vCJD and BSE. Nature 389:448–450

Holscher C, Delius H, Burkle A (1998) Overexpression of nonconvertible PrPc delta114–121 in scrapie-infected mouse neuroblastoma cells leads to trans-dominant inhibition of wild-type PrP(Sc) accumulation. J Virol 72:1153–1159

Hope J (1994) The nature of the scrapie agent: the evolution of the virino. Ann N Y Acad Sci 724:282–289

Hope J, Morton LJ, Farquhar CF, Multhaup G, Beyreuther K, Kimberlin RH (1986) The major polypeptide of scrapie-associated fibrils (SAF) has the same size, charge distribution and N-terminal protein sequence as predicted for the normal brain protein (PrP). EMBO J 5:2591–2597

Hope J, Multhaup G, Reekie LJ, Kimberlin RH, Beyreuther K (1988) Molecular pathology of scrapie-associated fibril protein (PrP) in mouse brain affected by the ME7 strain of scrapie. Eur J Biochem 172:271–277

Hope J, Wood SC, Birkett CR, Chong A, Bruce ME, Cairns D, Goldmann W, Hunter N, Bostock CJ (1999) Molecular analysis of ovine prion protein identifies similari-

ties between BSE and an experimental isolate of natural scrapie, CH1641. J Gen Virol 80 (Pt 1):1–4
Horiuchi M, Baron GS, Xiong LW, Caughey B (2001) Inhibition of interactions and interconversions of prion protein isoforms by peptide fragments from the C-terminal folded domain. J Biol Chem 276:15489–15497
Horiuchi M, Caughey B (1999a) Prion protein interconversions and the transmissible spongiform encephalopathies. Structure Fold Des 7:R231–R240
Horiuchi M, Caughey B (1999b) Specific binding of normal prion protein to the scrapie form via a localized domain initiates its conversion to the protease-resistant state. EMBO J 18:3193–3203
Horiuchi M, Nemoto T, Ishiguro N, Furuoka H, Mohri S, Shinagawa M (2002) Biological and biochemical characterization of sheep scrapie in Japan. J Clin Microbiol 40:3421–3426
Horiuchi M, Priola SA, Chabry J, Caughey B (2000) Interactions between heterologous forms of prion protein: binding, inhibition of conversion, and species barriers. Proc Natl Acad Sci USA 97:5836–5841
Hornemann S, Glockshuber R (1998) A scrapie-like unfolding intermediate of the prion protein domain PrP(121–231) induced by acidic pH. Proc Natl Acad Sci USA 95:6010–6014
Hornemann S, Korth C, Oesch B, Riek R, Wider G, Wuthrich K, Glockshuber R (1997) Recombinant full-length murine prion protein, mPrP(23–231): purification and spectroscopic characterization. FEBS Lett 413:277–281
Hornshaw MP, McDermott JR, Candy JM (1995) Copper binding to the N-terminal tandem repeat regions of mammalian and avian prion protein. Biochem Biophys Res Commun 207:621–629
Hunter N, Goldmann W, Smith G, Hope J (1994) Frequencies of PrP gene variants in healthy cattle and cattle with BSE in Scotland. Vet Rec 135:400–403
Jackson GS, Hosszu LL, Power A, Hill AF, Kenney J, Saibil H, Craven CJ, Waltho JP, Clarke AR, Collinge J (1999) Reversible conversion of monomeric human prion protein between native and fibrilogenic conformations. Science 283:1935–1937
Jackson GS, Murray I, Hosszu LL, Gibbs N, Waltho JP, Clarke AR, Collinge J (2001) Location and properties of metal-binding sites on the human prion protein. Proc Natl Acad Sci USA 98:8531–8535
Jarrett JT, Lansbury PT, Jr. (1993) Seeding "one-dimensional crystallization" of amyloid: a pathogenic mechanism in Alzheimer's disease and scrapie? Cell 73:1055–1058
Jeffrey M, Goodsir CM, Bruce ME, McBride PA, Farquhar C (1994) Morphogenesis of amyloid plaques in 87 V murine scrapie. Neuropathol Appl Neurobiol 20:535–542
Kaneko K, Peretz D, Pan KM, Blochberger TC, Wille H, Gabizon R, Griffith OH, Cohen FE, Baldwin MA, Prusiner SB (1995) Prion protein (PrP) synthetic peptides induce cellular PrP to acquire properties of the scrapie isoform. Proc Natl Acad Sci USA 92:11160–11164
Kaneko K, Vey M, Scott M, Pilkuhn S, Cohen FE, Prusiner SB (1997a) COOH-terminal sequence of the cellular prion protein directs subcellular trafficking and controls conversion into the scrapie isoform. Proc Natl Acad Sci USA 94:2333–2338

Kaneko K, Wille H, Mehlhorn I, Zhang H, Ball H, Cohen FE, Baldwin MA, Prusiner SB (1997b) Molecular properties of complexes formed between the prion protein and synthetic peptides. J Mol Biol 270:574–586

Kaneko K, Zulianello L, Scott M, Cooper CM, Wallace AC, James TL, Cohen FE, Prusiner SB (1997c) Evidence for protein X binding to a discontinuous epitope on the cellular prion protein during scrapie prion propagation. Proc Natl Acad Sci USA 94:10069–10074

Kascsak RJ, Rubenstein R, Carp RI (1991) Evidence for biological and structural diversity among scrapie strains. In: Chesebro B (eds) Transmissible Spongiform Encephalopathies: Scrapie, BSE and Related Human Disorders. Springer Verlag, Berlin, Heidelberg, pp 139–152

Kascsak RJ, Rubenstein R, Merz PA, Carp RI, Robakis NK, Wisniewski HM, Diringer H (1986) Immunological comparison of scrapie-associated fibrils isolated from animals infected with four different scrapie strains. J Virol 59:676–683

Kascsak RJ, Rubenstein R, Merz PA, Carp RI, Wisniewski HM, Diringer H (1985) Biochemical differences among scrapie-associated fibrils support the biological diversity of scrapie agents. J Gen Virol 66 (Pt 8):1715–1722

Knaus KJ, Morillas M, Swietnicki W, Malone M, Surewicz WK, Yee VC (2001) Crystal structure of the human prion protein reveals a mechanism for oligomerization. Nat Struct Biol 8:770–774

Kocisko DA, Come JH, Priola SA, Chesebro B, Raymond GJ, Lansbury PT, Caughey B (1994) Cell-free formation of protease-resistant prion protein. Nature 370:471–474

Kocisko DA, Lansbury PT, Jr., Caughey B (1996) Partial unfolding and refolding of scrapie-associated prion protein: evidence for a critical 16-kDa C-terminal domain. Biochemistry 35:13434–13442

Kocisko DA, Priola SA, Raymond GJ, Chesebro B, Lansbury PT, Jr., Caughey B (1995) Species specificity in the cell-free conversion of prion protein to protease-resistant forms: a model for the scrapie species barrier. Proc Natl Acad Sci USA 92:3923–3927

Korth C, Kaneko K, Prusiner SB (2000) Expression of unglycosylated mutated prion protein facilitates PrP(Sc) formation in neuroblastoma cells infected with different prion strains. J Gen Virol 81:2555–2563

Kryndushkin DS, Smirnov VN, Ter Avanesyan MD, Kushnirov VV (2002) Increased expression of Hsp40 chaperones, transcriptional factors, and ribosomal protein Rpp0 can cure yeast prions. J Biol Chem 277:23702–23708

Kuwahara C, Takeuchi AM, Nishimura T, Haraguchi K, Kubosaki A, Matsumoto Y, Saeki K, Matsumoto Y, Yokoyama T, Itohara S, Onodera T (1999) Prions prevent neuronal cell-line death. Nature 400:225–226

Lansbury PT, Jr., Caughey B (1995) The chemistry of scrapie infection: implications of the 'ice 9' metaphor. Chem Biol 2:1–5

Lawson VA, Priola SA, Wehrly K, Chesebro B (2001) N-terminal truncation of prion protein affects both formation and conformation of abnormal protease-resistant prion protein generated in vitro. J Biol Chem 276:35265–35271

Li R, Liu D, Zanusso G, Liu T, Fayen JD, Huang JH, Petersen RB, Gambetti P, Sy MS (2001) The expression and potential function of cellular prion protein in human lymphocytes. Cell Immunol 207:49–58

Liu H, Farr-Jones S, Ulyanov NB, Llinas M, Marqusee S, Groth D, Cohen FE, Prusiner SB, James TL (1999) Solution structure of Syrian hamster prion protein rPrP(90–231). Biochemistry 38:5362–5377

Locht C, Chesebro B, Race R, Keith JM (1986) Molecular cloning and complete sequence of prion protein cDNA from mouse brain infected with the scrapie agent. Proc Natl Acad Sci USA 83:6372–6376

Lucassen R, Nishina K, Supattapone S (2003) In vitro amplification of protease-resistant prion protein requires free sulfhydryl groups. Biochemistry 42:4127–4135

Ma J, Lindquist S (2002) Conversion of PrP to a self-perpetuating PrPSc-like conformation in the cytosol. Science 298:1785–1788

Marella M, Lehmann S, Grassi J, Chabry J (2002) Filipin prevents pathological prion protein accumulation by reducing endocytosis and inducing cellular PrP release. J Biol Chem 277:25457–25464

Masel J, Jansen VA, Nowak MA (1999) Quantifying the kinetic parameters of prion replication. Biophys Chem 77:139–152

Maxson L, Wong C, Herrmann LM, Caughey B, and Baron GS (2003) A solid-phase assay for identification of modulators of prion protein interactions. Anal Biochem (in press)

McKenzie D, Bartz J, Mirwald J, Olander D, Marsh R, Aiken J (1998) Reversibility of scrapie inactivation is enhanced by copper. J Biol Chem 273:25545–25547

McKinley MP, Meyer RK, Kenaga L, Rahbar F, Cotter R, Serban A, Prusiner SB (1991) Scrapie prion rod formation in vitro requires both detergent extraction and limited proteolysis. J Virol 65:1340–1351

Merz PA, Somerville RA, Wisniewski HM, Iqbal K (1981) Abnormal fibrils from scrapie-infected brain. Acta Neuropathol (Berl) 54:63–74

Miller MW, Williams ES, McCarty CW, Spraker TR, Kreeger TJ, Larsen CT, Thorne ET (2000) Epizootiology of chronic wasting disease in free-ranging cervids in Colorado and Wyoming. J Wildl Dis 36:676–690

Monari L, Chen SG, Brown P, Parchi P, Petersen RB, Mikol J, Gray F, Cortelli P, Montagna P, Ghetti B, . (1994) Fatal familial insomnia and familial Creutzfeldt–Jakob disease: different prion proteins determined by a DNA polymorphism. Proc Natl Acad Sci USA 91:2839–2842

Morillas M, Swietnicki W, Gambetti P, Surewicz WK (1999) Membrane environment alters the conformational structure of the recombinant human prion protein. J Biol Chem 274:36859–36865

Moriyama H, Edskes HK, Wickner RB (2000) [URE3] prion propagation in Saccharomyces cerevisiae: requirement for chaperone Hsp104 and curing by overexpressed chaperone Ydj1p. Mol Cell Biol 20:8916–8922

Morrissey MP, Shakhnovich EI (1999) Evidence for the role of PrP(C) helix 1 in the hydrophilic seeding of prion aggregates. Proc Natl Acad Sci USA 96:11293–11298

Mouillet-Richard S, Ermonval M, Chebassier C, Laplanche JL, Lehmann S, Launay JM, Kellermann O (2000) Signal transduction through prion protein. Science 289:1925–1928

Muramoto T, Scott M, Cohen FE, Prusiner SB (1996) Recombinant scrapie-like prion protein of 106 amino acids is soluble. Proc Natl Acad Sci USA 93:15457–15462

Nguyen JT, Inouye H, Baldwin MA, Fletterick RJ, Cohen FE, Prusiner SB, Kirschner DA (1995) X-ray diffraction of scrapie prion rods and PrP peptides. J Mol Biol 252:412–422

O'Rourke KI, Besser TE, Miller MW, Cline TF, Spraker TR, Jenny AL, Wild MA, Zebarth GL, Williams ES (1999) PrP genotypes of captive and free-ranging Rocky Mountain elk (Cervus elaphus nelsoni) with chronic wasting disease. J Gen Virol 80 (Pt 10):2765–2769

Oesch B, Jensen M, Nilsson P, Fogh J (1994) Properties of the scrapie prion protein: quantitative analysis of protease resistance. Biochemistry 33:5926–5931

Oesch B, Westaway D, Walchli M, McKinley MP, Kent SB, Aebersold R, Barry RA, Tempst P, Teplow DB, Hood LE (1985) A cellular gene encodes scrapie PrP 27–30 protein. Cell 40:735–746

Paitel E, Alves dC, Vilette D, Grassi J, Checler F (2002) Overexpression of PrPc triggers caspase 3 activation: potentiation by proteasome inhibitors and blockade by anti-PrP antibodies. J Neurochem 83:1208–1214

Paitel E, Fahraeus R, Checler F (2003) Cellular prion protein sensitizes neurons to apoptotic stimuli through Mdm2-regulated and p53-dependent caspase 3-like activation. J Biol Chem 278:10061–10066

Palmer MS, Dryden AJ, Hughes JT, Collinge J (1991) Homozygous prion protein genotype predisposes to sporadic Creutzfeldt–Jakob disease. Nature 352:340–342

Pan KM, Baldwin M, Nguyen J, Gasset M, Serban A, Groth D, Mehlhorn I, Huang Z, Fletterick RJ, Cohen FE (1993) Conversion of alpha-helices into beta-sheets features in the formation of the scrapie prion proteins. Proc Natl Acad Sci USA 90:10962–10966

Parchi P, Capellari S, Chen SG, Petersen RB, Gambetti P, Kopp N, Brown P, Kitamoto T, Tateishi J, Giese A, Kretzschmar H (1997) Typing prion isoforms. Nature 386:232–234

Parizek P, Roeckl C, Weber J, Flechsig E, Aguzzi A, Raeber AJ (2001) Similar turnover and shedding of the cellular prion protein in primary lymphoid and neuronal cells. J Biol Chem 276:44627–44632

Patino MM, Liu JJ, Glover JR, Lindquist S (1996) Support for the prion hypothesis for inheritance of a phenotypic trait in yeast. Science 273:622–626

Pauly PC, Harris DA (1998) Copper stimulates endocytosis of the prion protein. J Biol Chem 273:33107–33110

Peretz D, Scott MR, Groth D, Williamson RA, Burton DR, Cohen FE, Prusiner SB (2001a) Strain-specified relative conformational stability of the scrapie prion protein. Protein Sci 10:854–863

Peretz D, Williamson RA, Kaneko K, Vergara J, Leclerc E, Schmitt-Ulms G, Mehlhorn IR, Legname G, Wormald MR, Rudd PM, Dwek RA, Burton DR, Prusiner SB (2001b) Antibodies inhibit prion propagation and clear cell cultures of prion infectivity. Nature 412:739–743

Peretz D, Williamson RA, Legname G, Matsunaga Y, Vergara J, Burton DR, DeArmond SJ, Prusiner SB, Scott MR (2002) A change in the conformation of prions accompanies the emergence of a new prion strain. Neuron 34:921–932

Post K, Pitschke M, Schafer O, Wille H, Appel TR, Kirsch D, Mehlhorn I, Serban H, Prusiner SB, Riesner D (1998) Rapid acquisition of beta-sheet structure in the prion protein prior to multimer formation. Biol Chem 379:1307–1317

Priola SA (2001) Prion protein diversity and disease in the transmissible spongiform encephalopathies. Adv Protein Chem 57:1–27

Priola SA, Caughey B, Race RE, Chesebro B (1994) Heterologous PrP molecules interfere with accumulation of protease-resistant PrP in scrapie-infected murine neuroblastoma cells. J Virol 68:4873–4878

Priola SA, Chabry J, Chan K (2001) Efficient conversion of normal prion protein (PrP) by abnormal hamster PrP is determined by homology at amino acid residue 155. J Virol 75:4673–4680

Priola SA, Chesebro B (1995) A single hamster PrP amino acid blocks conversion to protease-resistant PrP in scrapie-infected mouse neuroblastoma cells. J Virol 69:7754–7758

Prusiner SB (1991) Molecular biology of prion diseases. Science 252:1515–1522

Prusiner SB (1998) Prions. Proc Natl Acad Sci USA 95:13363–13383

Prusiner SB, Groth D, Serban A, Stahl N, Gabizon R (1993) Attempts to restore scrapie prion infectivity after exposure to protein denaturants. Proc Natl Acad Sci USA 90:2793–2797

Prusiner SB, McKinley MP, Bowman KA, Bolton DC, Bendheim PE, Groth DF, Glenner GG (1983) Scrapie prions aggregate to form amyloid-like birefringent rods. Cell 35:349–358

Prusiner SB, Scott M, Foster D, Pan KM, Groth D, Mirenda C, Torchia M, Yang SL, Serban D, Carlson GA (1990) Transgenetic studies implicate interactions between homologous PrP isoforms in scrapie prion replication. Cell 63:673–686

Race R, Jenny A, Sutton D (1998) Scrapie infectivity and proteinase K-resistant prion protein in sheep placenta, brain, spleen, and lymph node: implications for transmission and antemortem diagnosis. J Infect Dis 178:949–953

Race RE, Raines A, Baron TG, Miller MW, Jenny A, Williams ES (2002) Comparison of abnormal prion protein glycoform patterns from transmissible spongiform encephalopathy agent-infected deer, elk, sheep, and cattle. J Virol 76:12365–12368

Raymond GJ, Bossers A, Raymond LD, O'Rourke KI, McHolland LE, Bryant PK, III, Miller MW, Williams ES, Smits M, Caughey B (2000) Evidence of a molecular barrier limiting susceptibility of humans, cattle and sheep to chronic wasting disease. EMBO J 19:4425–4430

Raymond GJ, Hope J, Kocisko DA, Priola SA, Raymond LD, Bossers A, Ironside J, Will RG, Chen SG, Petersen RB, Gambetti P, Rubenstein R, Smits MA, Lansbury PT, Jr., Caughey B (1997) Molecular assessment of the potential transmissibilities of BSE and scrapie to humans. Nature 388:285–288

Rieger R, Edenhofer F, Lasmezas CI, Weiss S (1997) The human 37-kDa laminin receptor precursor interacts with the prion protein in eukaryotic cells. Nat Med 3:1383–1388

Riek R, Hornemann S, Wider G, Billeter M, Glockshuber R, Wuthrich K (1996) NMR structure of the mouse prion protein domain PrP(121–321). Nature 382:180–182

Riek R, Hornemann S, Wider G, Glockshuber R, Wuthrich K (1997) NMR characterization of the full-length recombinant murine prion protein, mPrP(23–231). FEBS Lett 413:282–288

Riesner D, Kellings K, Post K, Wille H, Serban H, Groth D, Baldwin MA, Prusiner SB (1996) Disruption of prion rods generates 10-nm spherical particles having high alpha-helical content and lacking scrapie infectivity. J Virol 70:1714–1722

Rogers M, Yehiely F, Scott M, Prusiner SB (1993) Conversion of truncated and elongated prion proteins into the scrapie isoform in cultured cells. Proc Natl Acad Sci USA 90:3182–3186

Rubenstein R, Gray PC, Wehlburg CM, Wagner JS, Tisone GC (1998) Detection and discrimination of PrPSc by multi-spectral ultraviolet fluorescence. Biochem Biophys Res Commun 246:100–106

Rudd PM, Endo T, Colominas C, Groth D, Wheeler SF, Harvey DJ, Wormald MR, Serban H, Prusiner SB, Kobata A, Dwek RA (1999) Glycosylation differences between the normal and pathogenic prion protein isoforms. Proc Natl Acad Sci USA 96:13044–13049

Rydh A, Suhr O, Hietala SO, Ahlstrom KR, Pepys MB, Hawkins PN (1998) Serum amyloid P component scintigraphy in familial amyloid polyneuropathy: regression of visceral amyloid following liver transplantation. Eur J Nucl Med 25:709–713

Saborio GP, Permanne B, Soto C (2001) Sensitive detection of pathological prion protein by cyclic amplification of protein misfolding. Nature 411:810–813

Saborio GP, Soto C, Kascsak RJ, Levy E, Kascsak R, Harris DA, Frangione B (1999) Cell-lysate conversion of prion protein into its protease-resistant isoform suggests the participation of a cellular chaperone. Biochem Biophys Res Commun 258:470–475

Safar J, Roller PP, Gajdusek DC, Gibbs CJ, Jr. (1993a) Conformational transitions, dissociation, and unfolding of scrapie amyloid (prion) protein. J Biol Chem 268:20276–20284

Safar J, Roller PP, Gajdusek DC, Gibbs CJ, Jr. (1993b) Thermal stability and conformational transitions of scrapie amyloid (prion) protein correlate with infectivity. Protein Sci 2:2206–2216

Safar J, Roller PP, Gajdusek DC, Gibbs CJ, Jr. (1994) Scrapie amyloid (prion) protein has the conformational characteristics of an aggregated molten globule folding intermediate. Biochemistry 33:8375–8383

Safar J, Wille H, Itri V, Groth D, Serban H, Torchia M, Cohen FE, Prusiner SB (1998) Eight prion strains have PrP(Sc) molecules with different conformations. Nat Med 4:1157–1165

Safar JG, Scott M, Monaghan J, Deering C, Didorenko S, Vergara J, Ball H, Legname G, Leclerc E, Solforosi L, Serban H, Groth D, Burton DR, Prusiner SB, Williamson RA (2002) Measuring prions causing bovine spongiform encephalopathy or chronic wasting disease by immunoassays and transgenic mice. Nat Biotechnol 20:1147–1150

Sanghera N, Pinheiro TJ (2002) Binding of prion protein to lipid membranes and implications for prion conversion. J Mol Biol 315:1241–1256

Scott MR, Kohler R, Foster D, Prusiner SB (1992) Chimeric prion protein expression in cultured cells and transgenic mice. Protein Sci 1:986–997

Shaked GM, Fridlander G, Meiner Z, Taraboulos A, Gabizon R (1999) Protease-resistant and detergent-insoluble prion protein is not necessarily associated with prion infectivity. J Biol Chem 274:17981–17986

Shaked GM, Meiner Z, Avraham I, Taraboulos A, Gabizon R (2001a) Reconstitution of prion infectivity from solubilized protease-resistant PrP and nonprotein components of prion rods. J Biol Chem 276:14324–14328

Shaked GM, Shaked Y, Kariv-Inbal Z, Halimi M, Avraham I, Gabizon R (2001b) A protease-resistant prion protein isoform is present in urine of animals and humans affected with prion diseases. J Biol Chem 276:31479–31482
Shyng SL, Heuser JE, Harris DA (1994) A glycolipid-anchored prion protein is endocytosed via clathrin-coated pits. J Cell Biol 125:1239–1250
Shyng SL, Huber MT, Harris DA (1993) A prion protein cycles between the cell surface and an endocytic compartment in cultured neuroblastoma cells. J Biol Chem 268:15922–15928
Shyng SL, Lehmann S, Moulder KL, Harris DA (1995a) Sulfated glycans stimulate endocytosis of the cellular isoform of the prion protein, PrPC, in cultured cells. J Biol Chem 270:30221–30229
Shyng SL, Moulder KL, Lesko A, Harris DA (1995b) The N-terminal domain of a glycolipid-anchored prion protein is essential for its endocytosis via clathrin-coated pits. J Biol Chem 270:14793–14800
Sigurdsson EM, Permanne B, Soto C, Wisniewski T, Frangione B (2000) In vivo reversal of amyloid-beta lesions in rat brain. J Neuropathol Exp Neurol 59:11–17
Somerville RA, Chong A, Mulqueen OU, Birkett CR, Wood SC, Hope J (1997) Biochemical typing of scrapie strains. Nature 386:564
Somerville RA, Millson GC, Kimberlin RH (1980) Sensitivity of scrapie infectivity to detergents and 2-mercaptoethanol. Intervirology 13:126–129
Somerville RA, Ritchie LA (1990) Differential glycosylation of the protein (PrP) forming scrapie-associated fibrils. J Gen Virol 71 (Pt 4):833–839
Somerville RA, Ritchie LA, Gibson PH (1989) Structural and biochemical evidence that scrapie-associated fibrils assemble in vivo. J Gen Virol 70 (Pt 1):25–35
Soto C, Kascsak RJ, Saborio GP, Aucouturier P, Wisniewski T, Prelli F, Kascsak R, Mendez E, Harris DA, Ironside J, Tagliavini F, Carp RI, Frangione B (2000) Reversion of prion protein conformational changes by synthetic beta-sheet breaker peptides. Lancet 355:192–197
Speare JO, Rush TS, III, Bloom ME, Caughey B (2003) The role of helix 1 aspartates and salt bridges in the stability and conversion of prion protein. J Biol Chem 278:12522–12529
Spielhaupter C, Schatzl HM (2001) PrPC directly interacts with proteins involved in signaling pathways. J Biol Chem 276:44604–44612
Stahl N, Baldwin MA, Burlingame AL, Prusiner SB (1990a) Identification of glycoinositol phospholipid linked and truncated forms of the scrapie prion protein. Biochemistry 29:8879–8884
Stahl N, Borchelt DR, Hsiao K, Prusiner SB (1987) Scrapie prion protein contains a phosphatidylinositol glycolipid. Cell 51:229–240
Stahl N, Borchelt DR, Prusiner SB (1990b) Differential release of cellular and scrapie prion proteins from cellular membranes by phosphatidylinositol-specific phospholipase C. Biochemistry 29:5405–5412
Stimson E, Hope J, Chong A, Burlingame AL (1999) Site-specific characterization of the N-linked glycans of murine prion protein by high-performance liquid chromatography/electrospray mass spectrometry and exoglycosidase digestions. Biochemistry 38:4885–4895

Sunyach C, Jen A, Deng J, Fitzgerald KT, Frobert Y, Grassi J, McCaffrey MW, Morris R (2003) The mechanism of internalisation of glycosylphosphatidylinositol-anchored prion protein. EMBO J 22:3591–3601

Supattapone S, Bosque P, Muramoto T, Wille H, Aagaard C, Peretz D, Nguyen HO, Heinrich C, Torchia M, Safar J, Cohen FE, DeArmond SJ, Prusiner SB, Scott M (1999a) Prion protein of 106 residues creates an artifical transmission barrier for prion replication in transgenic mice. Cell 96:869–878

Supattapone S, Nguyen HO, Cohen FE, Prusiner SB, Scott MR (1999b) Elimination of prions by branched polyamines and implications for therapeutics. Proc Natl Acad Sci USA 96:14529–14534

Supattapone S, Wille H, Uyechi L, Safar J, Tremblay P, Szoka FC, Cohen FE, Prusiner SB, Scott MR (2001) Branched polyamines cure prion-infected neuroblastoma cells. J Virol 75:3453–3461

Swietnicki W, Petersen R, Gambetti P, Surewicz WK (1997) pH-dependent stability and conformation of the recombinant human prion protein PrP(90–231). J Biol Chem 272:27517–27520

Tagliavini F, Prelli F, Porro M, Rossi G, Giaccone G, Farlow MR, Dlouhy SR, Ghetti B, Bugiani O, Frangione B (1994) Amyloid fibrils in Gerstmann–Straussler–Scheinker disease (Indiana and Swedish kindreds) express only PrP peptides encoded by the mutant allele. Cell 79:695–703

Taraboulos A, Scott M, Semenov A, Avrahami D, Laszlo L, Prusiner SB, Avraham D (1995) Cholesterol depletion and modification of COOH-terminal targeting sequence of the prion protein inhibit formation of the scrapie isoform. J Cell Biol 129:121–132

Tatzelt J, Prusiner SB, Welch WJ (1996) Chemical chaperones interfere with the formation of scrapie prion protein. EMBO J 15:6363–6373

Telling GC, Parchi P, DeArmond SJ, Cortelli P, Montagna P, Gabizon R, Mastrianni J, Lugaresi E, Gambetti P, Prusiner SB (1996) Evidence for the conformation of the pathologic isoform of the prion protein enciphering and propagating prion diversity. Science 274:2079–2082

Thackray AM, Knight R, Haswell SJ, Bujdoso R, Brown DR (2002) Metal imbalance and compromised antioxidant function are early changes in prion disease. Biochem J 362:253–258

Tobler I, Deboer T, Fischer M (1997) Sleep and sleep regulation in normal and prion protein-deficient mice. J Neurosci 17:1869–1879

Tobler I, Gaus SE, Deboer T, Achermann P, Fischer M, Rulicke T, Moser M, Oesch B, McBride PA, Manson JC (1996) Altered circadian activity rhythms and sleep in mice devoid of prion protein. Nature 380:639–642

Tuo W, O'Rourke KI, Zhuang D, Cheevers WP, Spraker TR, Knowles DP (2002) Pregnancy status and fetal prion genetics determine PrPSc accumulation in placentomes of scrapie-infected sheep. Proc Natl Acad Sci USA 99:6310–6315

Turk E, Teplow DB, Hood LE, Prusiner SB (1988) Purification and properties of the cellular and scrapie hamster prion proteins. Eur J Biochem 176:21–30

Tzaban S, Friedlander G, Schonberger O, Horonchik L, Yedidia Y, Shaked G, Gabizon R, Taraboulos A (2002) Protease-sensitive scrapie prion protein in aggregates of heterogeneous sizes. Biochemistry 41:12868–12875

Viles JH, Cohen FE, Prusiner SB, Goodin DB, Wright PE, Dyson HJ (1999) Copper binding to the prion protein: structural implications of four identical cooperative binding sites. Proc Natl Acad Sci USA 96:2042–2047
Vorberg I, Chan K, Priola SA (2001) Deletion of beta-strand and alpha-helix secondary structure in normal prion protein inhibits formation of its protease-resistant isoform. J Virol 75:10024–10032
Vorberg I, Priola SA (2002) Molecular basis of scrapie strain glycoform variation. J Biol Chem 277:36775–36781
Wadsworth JD, Hill AF, Joiner S, Jackson GS, Clarke AR, Collinge J (1999) Strain-specific prion-protein conformation determined by metal ions. Nat Cell Biol 1:55–59
Waggoner DJ, Drisaldi B, Bartnikas TB, Casareno RL, Prohaska JR, Gitlin JD, Harris DA (2000) Brain copper content and cuproenzyme activity do not vary with prion protein expression level. J Biol Chem 275:7455–7458
Weissmann C (1991) A 'unified theory' of prion propagation. Nature 352:679–683
Weissmann C (1999) Molecular genetics of transmissible spongiform encephalopathies. J Biol Chem 274:3–6
Welker E, Raymond LD, Scheraga HA, Caughey B (2002) Intramolecular versus intermolecular disulfide bonds in prion proteins. J Biol Chem 277:33477–33481
Welker E, Wedemeyer WJ, Scheraga HA (2001) A role for intermolecular disulfide bonds in prion diseases? Proc Natl Acad Sci USA 98:4334–4336
White AR, Enever P, Tayebi M, Mushens R, Linehan J, Brandner S, Anstee D, Collinge J, Hawke S (2003) Monoclonal antibodies inhibit prion replication and delay the development of prion disease. Nature 422:80–83
Will RG, Ironside JW, Zeidler M, Cousens SN, Estibeiro K, Alperovitch A, Poser S, Pocchiari M, Hofman A, Smith PG (1996) A new variant of Creutzfeldt-Jakob disease in the UK. Lancet 347:921–925
Wille H, Michelitsch MD, Guenebaut V, Supattapone S, Serban A, Cohen FE, Agard DA, Prusiner SB (2002) Structural studies of the scrapie prion protein by electron crystallography. Proc Natl Acad Sci USA 99:3563–3568
Wille H, Prusiner SB (1999) Ultrastructural studies on scrapie prion protein crystals obtained from reverse micellar solutions. Biophys J 76:1048–1062
Williams ES, Miller MW (2002) Chronic wasting disease in deer and elk in North America. Rev Sci Tech 21:305–316
Winklhofer KF, Tatzelt J (2000) Cationic lipopolyamines induce degradation of PrPSc in scrapie-infected mouse neuroblastoma cells. Biol Chem 381:463–469
Wong BS, Chen SG, Colucci M, Xie Z, Pan T, Liu T, Li R, Gambetti P, Sy MS, Brown DR (2001a) Aberrant metal binding by prion protein in human prion disease. J Neurochem 78:1400–1408
Wong C, Xiong LW, Horiuchi M, Raymond L, Wehrly K, Chesebro B, Caughey B (2001b) Sulfated glycans and elevated temperature stimulate PrPSc-dependent cell-free formation of protease-resistant prion protein. EMBO J 20:377–386
Wuthrich K, Riek R (2001) Three-dimensional structures of prion proteins. Adv Protein Chem 57:55–82
Xiong LW, Raymond LD, Hayes SF, Raymond GJ, Caughey B (2001) Conformational change, aggregation and fibril formation induced by detergent treatments of cellular prion protein. J Neurochem 79:669–678

Zanata SM, Lopes MH, Mercadante AF, Hajj GN, Chiarini LB, Nomizo R, Freitas AR, Cabral AL, Lee KS, Juliano MA, de Oliveira E, Jachieri SG, Burlingame A, Huang L, Linden R, Brentani RR, Martins VR (2002) Stress-inducible protein 1 is a cell surface ligand for cellular prion that triggers neuroprotection. EMBO J 21:3307–3316

Zeidler M, Stewart G, Cousens SN, Estibeiro K, Will RG (1997) Codon 129 genotype and new variant CJD. Lancet 350:668

Zhang H, Stockel J, Mehlhorn I, Groth D, Baldwin MA, Prusiner SB, James TL, Cohen FE (1997) Physical studies of conformational plasticity in a recombinant prion protein. Biochemistry 36:3543–3553

Zou WQ, Cashman NR (2002) Acidic pH and detergents enhance in vitro conversion of human brain PrPC to a PrPSc-like form. J Biol Chem 277:43942–43947

Past, Present and Future of Bovine Spongiform Encephalopathy in France

D. Calavas[1] · C. Ducrot[2] · T. G. M. Baron[1]

[1] Unité Epidémiologie, AFSSA Lyon, 31 av. Tony Garnier, 69364, Lyon cedex 07, France
E-mail: d.calavas@lyon.afssa.fr

[2] Unité d'Epidémiologie Animale, INRA Theix, 63122 , St. Genès Champanelle, France

Abstract The bovine spongiform encephalopathy (BSE) epidemic has been monitored in France since the end of 1990. The surveillance has been considerably enhanced since 2000, and today every cow aged 2 years or more is tested at the time of slaughter, culling or death. As of 1 May 2002, 613 native cases have been identified, 287 of them by the mandatory reporting system of suspect clinical cases or complementary programs, 213 by active surveillance of fallen stock and 113 by testing at the abattoir. The analysis of reported BSE cases shows a higher number of cases born between 1993 and 1995, which can be linked to a greater exposure at that time and to an increase in surveillance efficiency. When the clinical onset related to overexposure ends, the future trend of the BSE epidemic in France will depend on the efficiency of the control measures implemented since 1996. An indicator of this will be the number of BSE cases born among recent cohorts.

Bovine spongiform encephalopathy (BSE) was described first in 1987 in Great Britain (Wells et al. 1987), where a huge epidemic started, and more than 180,000 cases have been detected since then in this country. A risk analysis undertaken in France in 1990, in part because of the amount of meat and bone meal (MBM) imported from Great Britain during the 1980s, led to the conclusion that BSE might have spread to France, and that sporadic cases might be observed (Savey et al. 1991). Epidemiological surveillance was therefore set up at the end of 1990, and control measures were taken to prevent the development of the disease. The trends in the number of BSE cases detected in France must be analyzed in the light of both the detection system and the control measures, as well as their changes over time.

1 Evolution of BSE Surveillance

BSE was declared a notifiable disease on 13 June 1990, and a mandatory reporting system was set up. The surveillance system improved progressively over time, allowing us to achieve today an efficient surveillance of the disease. Two main steps in the surveillance can be identified, the first based only on clinical surveillance, the second on a combination of clinical surveillance and test programs targeted at certain bovine subpopulations.

1.1 Clinical Surveillance: December 1990–May 2000

The mandatory reporting system in force since the end of 1990 is a so-called 'passive' system, in the sense that it is based neither on systematic biological tests nor on targeting certain bovine subpopulations; rather, it consists in the identification of suspect clinical cases through surveillance of the whole adult bovine population (aged 2 years or more), on the basis of clinical, epidemiological and anamnetic criteria. Veterinarians declare as suspect the animals that display clinical signs including alterations in behavior, locomotion, or hypersensitivity, leading progressively to death. Suspect cases are entered into the mandatory reporting system of suspect clinical cases found either on farms by the veterinarian or at the abattoir at the time of antemortem examination.

Complementary programs have been undertaken over time. Since the end of 1998, animals suspected of rabies are systematically checked for BSE. Since May 1999, the surveillance of emergency-slaughtered animals (particularly those showing alterations in behavior and locomotion) has been strengthened, and imported cows from Switzerland and Portugal (two countries where the incidence of BSE is higher than in France) have been controlled at culling. In November, 1999, following an EU regulation (directive 98–272), a complementary program was launched, in order to reach a given number of BSE analyses on slaughtered animals, focusing on those showing a progressive disease syndrome or a poor body condition. In 2000, an experimental clinical network was started based on sentinel veterinarians, intended to give an estimate of the incidence and distribution of neurological diseases other than BSE (Calavas et al. 2001a).

In all these programs, after euthanasia, the diagnosis was confirmed by the identification of specific spongiform lesions in the brain stem of the animal, or, when histology was not possible, by the detection of protease resistant prion (PrP^{res}) by Western blot.

1.2
Clinical Surveillance and Test Programs: (Since June 2000)

A pilot screening program using rapid tests for the detection of PrP^{res} in BSE infected animals was set up for the first time in France in June 2000 to get more precise estimates of the BSE epidemic in France (Calavas et al. 2001b; Morignat 2001). It was implemented in the three regions of North-Western France considered to be the most affected by BSE (according to the mandatory reporting system): Basse Normandie, Bretagne and Pays de la Loire. This program started in June 2000 and ended in March 2001. Three categories of cattle considered to be at risk for BSE (Doherr et al. 1999) (dead on-farm, subjected to euthanasia, and emergency slaughtered) were sampled exhaustively. The samples were checked using the Prionics test (Moynagh et al. 1999), and samples that did not appear negative by this test were checked by Western blot or/and immunohistochemistry. In the same period, a complementary program was undertaken between November and December 2000 in the rest of the country on a sample of the same categories of animals at risk, following the EU regulation no. 2000/374/CE. The surveillance of the fallen stock (cows aged 2 years or more) resumed on 18 June 2001 in all three regions for at least 1 year, following another EU regulation. Prionics ND or Biorad ND tests are used for this program that is currently running.

Last but not least, since January 2001 every cow aged 30 months or more is tested at slaughter (EU regulation 2000/764/CE), using either Prionics ND or Biorad ND tests. The minimum age has been lowered to 24 months since July 2001, in order to be consistent with the overall BSE surveillance system, namely the mandatory reporting system and fallen stock surveillance.

2
Control Measures of BSE

Control measures have been progressively modified in France, according to new scientific knowledge. A major change followed the demonstration that BSE was transmissible to humans, when the emergence of a 'variant' form of Creutzfeldt–Jakob disease (vCJD), linked to BSE, was discovered in 1996.

The main control measures concerning animal feed are the following:

1. A ban on the use of MBM, which was shown, through epidemiological studies performed in Great Britain at the very beginning of the epidemic, to be the main contamination source in cattle
2. The removal from MBM processing of dead animals and 'Specified Risk Materials' (SRM) which are all the animal tissues that contain, or may contain, the BSE agent, mainly cattle brain and spinal cord

The main control measures are listed in Table 1 together with the date at which these measures started. From these dates, different categories of BSE cases can be distinguished according to their putative origin. Also as described in Table 1, birth of animals before or after each of the main control measures can give some information about the possible source of exposure to the BSE agent at the time.

The major measure to control MBM contamination, taken in 1996 in France, which is the removal of cadavers and SRM (specified risk material), has also been reinforced on several occasions. Measures were taken in 1998 to increase the temperature at which MBM had to be heated. In addition, some tissues which may contain low doses of infectious agent have been added to the list of specified risk material, especially intestines and vertebral columns, that now must be removed from the

Table 1 Synopsis of BSE control measures taken in France and definition of the different types of BSE cases in relation to enforcement of the main control measures

Date	BSE control measures	Types of BSE cases
		↓
June 1989	Ban on the import from Great Britain of live cattle (over 6 months) and of meat-and-bone meal	'Born before the ban' (BBB) cases
		↓
July 1990	Ban on meat-and-bone meal for cattle feed	'Born after the ban' (BAB) cases
		↓
December 1994	Ban on meat-and-bone meal for all ruminant feed	'Born after the ban' (BAB) cases
		↓
August 1996	Removal of specified risk material and cadavers from meat-and-bone meal	'Born after the second ban' (BASB) cases
		↓
November 2000	Ban on meat-and-bone meal and on certain animal fats for all farmed animals	'Born after the total ban' (BATB) cases
		↓
		↓

food chain. Finally these measures have been completed by the ban of MBM and of certain animal fats for all farmed animals in November 2000.

It should be emphasized, however, that control measures such as the removal of SRM and the ban on MBM were not set up at the same time in other European countries. In Great Britain, the ban on MBM was implemented in July 1988, and the removal of specified risk material in 1990. In contrast, such measures were not taken in some other European countries until 2000, including countries neighboring France, in which the first BSE cases were discovered recently, following the systematic use of rapid tests at the slaughterhouse or/and in rendering plants. The removal of cadavers and SRM only became a regulation at the European level in October 2000. This means that legally imported MBM did not necessarily have the same security level than those of French origin. The risk of accidental cross-contamination or illegal import of forbidden MBM also existed beyond 1996.

3
Epidemiological Data

The data from the different programs of the French BSE surveillance system are summarized in Table 2.

Up to 1 May 2002, 613 native BSE cases and one case imported from Switzerland had been detected. The mandatory reporting system detected roughly half of the cases (46.9%), the fallen stock programs 34.7%, and tests at the abattoir 18.4%.

The geographical distribution of the cases (Fig. 1) must be analyzed with caution because the surveillance was carried out differently depending on the regions and periods: the pilot program, in particular, that detected 70 cases on fallen stock between June 2000 and March 2001, was carried out in only three regions of Western France that represent one-third of the French bovine population.

Of the 613 native cases, 27 are BBB cases, 559 are BAB cases and 20 are BASB cases of which 10 were born more than 4 months after the im-

Table 2 Number of tested animals and positive native BSE cases, depending on the BSE surveillance programs and periods, from 1 December 1990 to 1 May 2002 (NA, not available)

Program	Specifications	Period	Tested (*n*)	Positive (*n*)
Passive surveillance	Adult bovine 24 months and over	Since 1 December 1990	1,021	281
Complementary programs	Rabies suspicions Enhanced surveillance in abattoir Imported animals from Switzerland and Portugal Poor body condition slaughtered animals Experimental network on nervous diseases	1998–2000	NA	6
Fallen stock programs	Pilot study in western France	June 2000–March 2001	44,464	70
	Complementary program in other regions	November–December 2000	5,400	6
	Census	Since 18 June 2001	216,552	137
Abattoir program	Slaughtered animals 30 months and over (24 months since July 2001)	Since January 2001	3,348,174	113

Fig. 1 Geographical distribution of BSE cases in France detected between 1 December 1990 and 1 May 2002

plementation of the ban. The month of birth of seven cases is not known; it is therefore impossible to classify them according to the different BSE case types. No BATB (animals born after November 2000) case has been detected so far but this can be explained by the minimum putative incubation time of the disease.

4 Analysis of the Trends of the BSE Epidemic in France

4.1 The Past

The efficiency of a clinical surveillance system depends on many parameters including technical, economical and sociological ones, leading to underidentification and underdeclaration of clinically suspect cases.

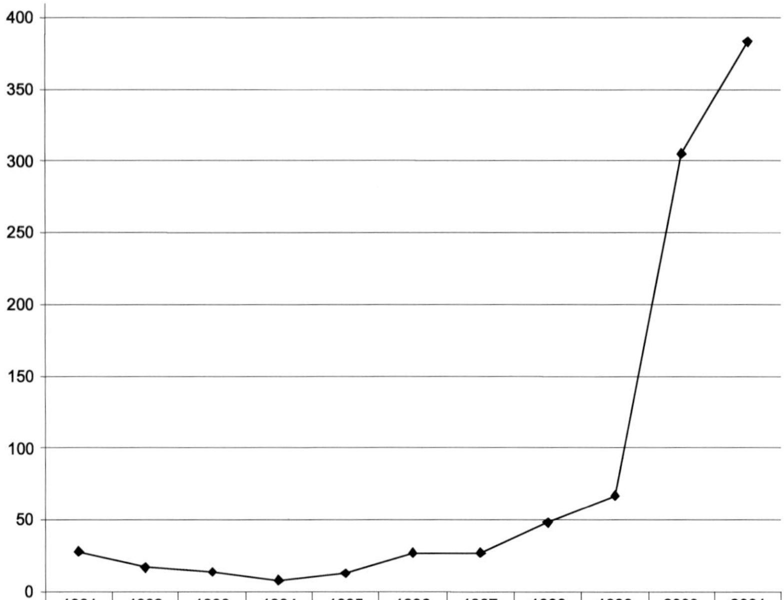

	1991	1992	1993	1994	1995	1996	1997	1998	1999	2000	2001
Nb	28	17	14	8	13	27	27	48	67	305	383

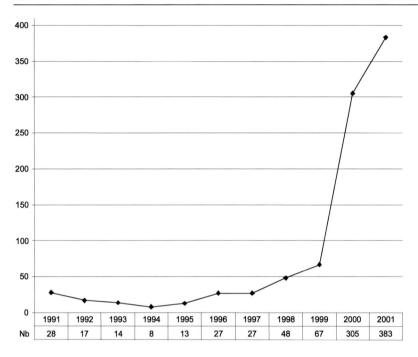

Fig. 2 Distribution of negative suspicions in the BSE mandatory reporting system in France between 1991 and 2001

Concurrent with successive improvements in the surveillance system, it is possible that the awareness of farmers and veterinarians varied over time along with scientific knowledge concerning the link between BSE and vCJD, the number of cases detected in France, media coverage and the adoption in 2000 of active surveillance programs. An indicator of the variation in clinical surveillance efficiency over time is the number of clinically suspect cases each year—more precisely the number of negative suspicions; this can be used to distinguish the indicator from the trend of the epidemic itself (Fig. 2). This indicator shows two small spikes, at the beginning of the surveillance and in 1996—the year of the first 'mad cow' crisis—then a slow increase between 1997 and 1999, and finally a huge increase, the number of negative suspicions in 2001 being 5.7 times higher than in 1999. Until 1999 the number of clinical suspicions in France has been low in relation to the size of the bovine population. As an example, during the 1990s, the number of clinically suspected cases normalized to the bovine population was 17 times lower in France than in Switzerland. As early as 1993, the limits of the surveillance system have been underlined (Coudert et al. 1995). Therefore, it is necessary to take into account the under identification/declaration of the cases to obtain an estimate of the real size of the BSE epidemic in France. This was done by mathematical modeling in two studies based on the number of cases identified by the clinical surveillance system from 1991 to the end of 2000; the first one is a back-calculation model (Donnelly 2000), and the second uses an age–period–cohort model (Cohen and Calavas 2002). According to Donelly's results, between 4700 and 9800 cows could have been infected in France, born mainly in the late 1980s. So it appears that several hundred cases might not have been identified, with a peak in 1993 or 1994 (Cohen and Calavas 2002), before control measures taken to break the recycling of the BSE agent were in force.

In these two studies the level of declaration was arbitrarily set to 100% in 2000. The active surveillance programs, implemented a part of the country since 2000 and in the whole country since mid-2001, give an estimate of the real declaration level for these years that should be taken into account to update the models. For example the estimates would be twice as high if the true declaration level was set to 50% in 2000 instead of to 100%.

4.2 The Present

Most of the BSE cases detected in France were born between 1993 and 1995 (Fig 3). The same trend is observed regardless of the detection method (clinical surveillance or targeted surveillance with rapid tests). The higher figures for cohorts in 1993–1995 are partly explained by the increase in the surveillance efficiency that came in force in 2000 (as cohorts born between 1993 and 1995 would then be at the age at which clinical signs would be apparent)

Given the hypothesis of a preferential contamination during the first year of life, we can postulate that the exposure level was high between 1993 and 1995. Different explanations can be given. Firstly, the recycling of MBM from affected animals in the first BSE wave, i.e., cows born and contaminated in 1988–1989 that were clinically diseased and that died between 1993 and 1995 without being detected as BSE cases and hence

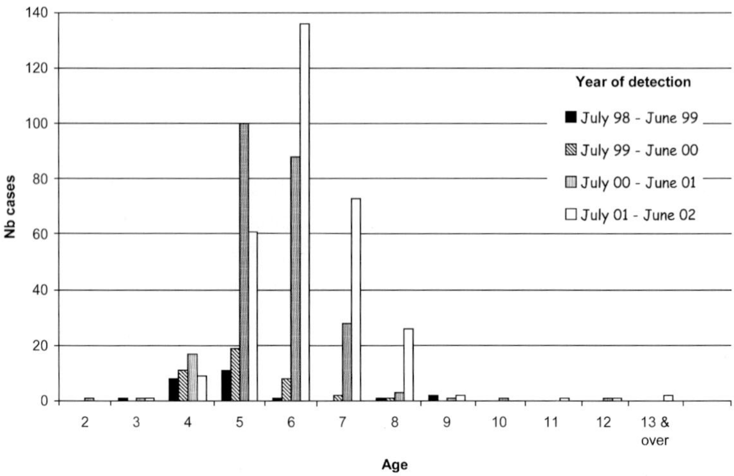

Fig. 3 Distribution of the age of BSE cases in France, at the time of detection, according to the year of detection

were not removed from the MBM industry. This is in agreement with results obtained from models (Donnelly 2000; Cohen and Calavas 2002). A second explanation is the import of MBM from European countries, other than Great Britain, with the same potential contamination of MBM with affected animals not recognized as BSE cases. The third explanation is the illegal import of MBM from Great Britain.

Because the use of MBM has been banned for cattle since 1990 in France, the previous explanations are valid if we make the assumption that either MBM was used illegally in cattle feed, or there was cross-contamination between monogastric feed—for which MBM was still authorized—and cattle feed; this cross-contamination might have happened during feed processing, during the shipment, or on the farm. Although this cross-contamination hypothesis seems the most likely, another one has to be taken into consideration. It is the use of authorized products derived from cattle carcasses, like tallow and dicalcic phosphates issued from bones. These different hypotheses are currently under investigation in an epidemiological study including a case control study and a geographical analysis.

4.3 The Future

Given current estimates of the incubation time, and analysis of data on previous cohorts, most clinical cases of a given cohort arise between 4 and 7 years of age; that means between 1997 and 2000 for the 1993 cohort, between 1998 and 2001 for the 1994 cohort, etc. Hence, the contamination period between 1993 and the first half of 1996, in conjunction with the increase in the surveillance efficiency since 2000, should induce a peak in the number of BSE cases in 2000 and 2001, then a decrease of the epidemic during the following 3 years.

Apart from these cases, the future trend of the epidemic in France will depend mostly on the efficiency of control measures started in 1996, which involves three major aspects.

The first question is whether or not the reinforced control measures introduced in 1996 were applied efficiently to the whole country. The poor results from the 1990 feed ban require caution; the emergence of BASB cases since 1996 showed that the feed ban was not sufficient and/or had not been applied carefully enough to stop the contamination process. In an opinion on BSE risk related to MBM and tallow, in April 2001, the French Agency for Food Safety (AFSSA) mentions that 'none of the key steps to make the process secure should be considered 100% effective, so that we could consider the MBM safe' (AFSSA 2001).

The second question relates to the risk resulting from illegal MBM imports from Great Britain and from legal imports from other European countries that did not apply the ban on cadavers and SRM from the MBM process in 1996. As the AFSSA stated in the same opinion 'it seems likely that the countries that did not require for their own regulation, the ban of cadavers and risk materials, could hardly guarantee that exported MBM was issued by distinct processing' (AFSSA 2001). Depending on which countries were the source of the annual 30,000 imported tons, and if cross-contamination between monogastric and ruminant feed were not controlled (in the plant, during the shipment, or on the farm), there was a probable source of contamination of cows born after 1996.

The third question is whether or not current control measures take into account the whole route of transmission of the disease. Do other transmission routes exist that are not currently identified, that might allow the epidemic to be maintained in cattle? British data show that if other transmission routes do exist, their role in transmission dynamics might be small, but they could maintain BSE in cattle at a low level. The answer to these three questions is not known yet, but the efficiency of reinforced control measures in June 1996 is under analysis already, thanks to the surveillance system based both on clinical signs and on screening programs. On 1 May 2002, 20 BASB cases (born after the reinforced control measures and the ban on cadavers and risk materials in MBM) were detected, and 10 of those cases were born more than 4 months after the implementation of these measures; this demonstrates that failures still persist in the control system. However, analysis of the age distribution of BSE cases detected during each of the last 4 years (Fig. 3) shows that the relative proportion of animals born after the reinforced ban decreases; this is the case for the 4-year-old animals among cases detected from July 2000 to June 2001, as well as for the 4- and 5-year-old animals detected since July 2001. This trend is difficult to interpret since it is based on surveillance data that changed over time (which explains the huge differences in the number of cases detected per year in Fig. 3). Further quantitative analyses are under way.

Finally the future of the BSE epidemic within the next 3 years will also depend on the complementary control measures taken at the European level in November 2000, based on the total ban of MBM and tallow for farmed animals.

5 Conclusion

The evolution of the French BSE surveillance apparatus shows that a great proportion of cases was probably not detected among cohorts born in the late 1980s; this is in agreement with back calculation and age–year–period models built on the existing data. Now, we may have a more optimistic view of the coming evolution of the BSE epidemic. The dramatic increase in the number of cases since 2000 is a consequence both of a huge increase in surveillance efficiency—due to the use of the rapid tests on both dead and slaughtered animals—and the pike in clinical cases linked to contamination that took place between 1993 and 1996.

When this BSE pike is over, presumably in 2002, we might observe a decrease in BSE incidence due to the reinforced control measures taken in 1996. The preliminary analysis of the age distribution of cases detected during each of the last 4 years shows a decrease in the trend of the epidemic that requires deeper investigation. The efficiency of the reinforced control measures will depend on their effectiveness at the abattoir, the risk due to the MBM import from Great Britain or other countries that did not control the contamination risk of MBM, and the presence of other contamination sources.

Since 2001, France as well as most other European countries have a reliable and comparable detection system, with clinical surveillance and active surveillance on dead and slaughtered animals as complementary tools. This will allow soon the comparative analysis of the prevalence rate of BSE in different age cohorts between countries.

References

AFSSA. Les risques sanitaires liés aux différents usages des farines et graisses d'origine animale et aux conditions de leur traitement et de leur élimination. 2001: Maisons-Alfort. 200 pp

Calavas D, Desjouis G, Collin E, Schelcher F, Philippe S, Savey M. Incidence et typologie des maladies des bovins adultes à expression nerveuse en France. Le réseau pilote NBA. Epidémiologie et Santé Animale, 2001a; 39:121–129

Calavas D, Ducrot C, Baron T, et al. Prevalence of BSE in western France by screening cattle at risk: preliminary results of a pilot study. The Veterinary Record, 2001b; 149(2): 55–56

Cohen CH, Calavas D. Analyse des variations temporelles de l'incidence de l'ESB en France. in Journées scientifiques de l' AFSSA. 2002. Maisons-Alfort, 27–28 March 2002

Coudert M, Belli P, Savey M, Martel J-L. Le réseau national d'épidémiosurveillance de l'encéphalopathie spongiforme bovine. Epidémiologie et Santé Animale, 1995; 27:59–67

Doherr MG, Oesch B, Moser M, Vandevelde M, Heim D. Targeted surveillance for bovine spongiform encephalopathy. The Veterinary Record, 1999; 145:672

Donnelly CA. Likely size of the French BSE epidemic. Nature, 2000; 408:787–788

Morignat E, Ducrot C, Roy P, et al. Targeted surveillance to assess the prevalence of BSE in high risk populations in western France and associated risk factors. The Veterinary Record, 2001; 151(3):73–77

Moynagh J, Schimmel H, Kramer G. The evaluation of tests for the diagnosis of transmissible spongiform encephalopathy in bovines. 1999, European Commission

Savey M, Belli P, Coudert M. Le réseau d'épidémiosurveillance de la BSE en France: principes — premiers résultats. Epidémiologie et Santé Animale 1991; 19:49–61

Wells GAH, Scott AC, Johnson CT, et al. A novel progressive spongiform encephalopathy in cattle. The Veterinary Record 1987; 121:419–420

Pathology and Pathogenesis of Bovine Spongiform Encephalopathy and Scrapie

M. Jeffrey · L. González
Veterinary Laboratories Agency (VLA-Lasswade), Pentlands Science Park, Bush Loan, Penicuik, Midlothian, EH26 0PZ, UK
E-mail: M.Jeffrey@vla.defra.gsi.gov.uk

Abstract In common with other prion diseases or transmissible spongiform encephalopathies (TSEs), scrapie of sheep and bovine spongiform encephalopathy (BSE) are characterized by grey matter vacuolation and accumulation of an abnormal isoform of the host prion protein (PrP) in the central nervous system (CNS). In apparent contrast with human disease, neither neuronal loss nor gliosis are invariable features of the pathology of domestic food animal TSEs. In sheep, accumulation of abnormal PrP may also occur in the lymphoreticular and peripheral nervous systems where it may be detected within months of birth. The involvement of tissues other than CNS is influenced by dose, PrP genotype of the host and strain of TSE agent. Although many different strains of scrapie agent have been isolated in rodents following serial passage of affected sheep brain tissue, the significance of these murine strains for natural sheep scrapie, and the extent to which different sheep scrapie strains occur naturally are uncertain. Whereas the consistent vacuolar pattern in the brains of BSE-affected cattle suggests a single strain of agent, the patterns of vacuolation in sheep scrapie are highly variable and cannot be easily used to define strain. In sheep scrapie, immunohistochemistry can be used to visualize different morphological types of abnormal PrP within individual brains. These different types of PrP accumulation seem to be associated with different brain cell types and with variation in the processing of abnormal PrP. When assessed in whole brain, different patterns of PrP accumulation are helpful in distinguishing between different sheep scrapie strains and also between ovine BSE and natural sheep scrapie.

1 Introduction

Scrapie is considered the archetypal transmissible spongiform encephalopathy (TSE). It is a fatal neurodegenerative disease of sheep and goats that has a widespread geographical distribution and has been present in Europe for centuries. In contrast, bovine spongiform encephalopathy (BSE) is a related TSE disorder that was first recognized in Great Britain affecting cattle (Wells et al. 1987) and exotic ruminants (Jeffrey and Wells 1988). The contemporary occurrence of TSEs in domestic cats (Wyatt et al. 1990), in a wide range of exotic ungulates (Cunningham et al. 1993) and felids kept in zoological collections and wildlife parks (Willoughby et al. 1992) was initially attributed also to infection with the BSE agent; this has subsequently been confirmed on the transmission characteristics of these diseases in mice (Bruce et al. 1994).

It is generally considered that BSE arose from sheep scrapie contaminated material entering the cattle food chain (Wilesmith et al. 1991). The BSE agent has been experimentally transmitted to a number of domestic species and laboratory animals and, in common with other TSEs, infection results in a host specific pathological response (Wells and McGill 1992; Nathanson et al. 1999). This review will describe the pathology of natural and experimental BSE in cattle, experimental BSE in sheep and natural and experimental scrapie of sheep.

The nature of the infectious agent of TSEs remains controversial but a majority view would support the prion hypothesis (Prusiner 1982) according to which the agent would be a 'proteinaceous infectious particle that lacks nucleic acid; composed largely, if not entirely, of PrP^{sc} molecules' (Prusiner 1999b). PrP^{sc} is defined as an 'abnormal, pathogenic isoform of the prion protein that causes sickness; the only identifiable macromolecule in purified preparations of prions'. The abnormal form of the prion protein is derived from a host coded cell surface sialoglycoprotein designated as PrP^{c}. In many studies, the abnormal form of the prion protein has been shown to possess a number of qualities that differentiate it from normal PrP^{c} (Prusiner 1999a). These properties include partial protease resistance, insolubility and high β-pleated sheet content (Prusiner 1999a). However, it is also stated that these properties should only be used as 'surrogate markers of PrP^{sc}, since these properties may not always be present' (Prusiner 1999b). Therefore, the precise molecular and biochemical characteristics of PrP^{sc} still need to be accurately defined. Because of this lack of a clear definition of the term PrP^{sc} and its relationship to the hypothetical prion, in this review we will use the designation PrP^{d} to indicate the disease-specific accumulation of PrP detected in immunohistochemical preparations, without prejudice as to its solubility, protease resistance or β-pleated sheet content. PrP^{res} will be used to indicate the disease-specific, partially proteinase-resistant fraction of PrP obtained after enzymatic digestion of infected tissue samples.

A long standing problem for the prion hypothesis has been how to explain the many different laboratory strains of scrapie that have been characterized in mice (Bruce et al. 1992) and that are increasingly being recognized in sheep (Bruce et al. 2002; Hope et al. 1999; González et al. 2002). More than 20 strains of scrapie have been defined in mice, which, according to the prion hypothesis, indicate that numerous stable conformational differences in the tertiary or quaternary structure of PrP^{sc} are capable of producing strain specific patterns of pathology and incubation period. Although some suggestions have been made to modify the prion hypothesis taking account of the difficulties of explaining strain variation within this hypothesis (Weissmann 1992), a minority view still maintains that a conventional nucleic acid underpins strain diversity (Farquhar et al. 1998). In the present review we will also describe features of sheep TSEs which indicate that different abnormal forms of PrP may be found after infections with different TSE sources or agents. In addition, however, multiple abnormal forms of PrP appear to be produced by different cells following infection with the same TSE agent, suggesting that structural variations of the abnormal form of PrP does not invariably code for different scrapie strains.

2 Histopathology

2.1 Central Nervous System Changes in BSE-Affected Cattle

The initial pathological case definition of BSE was based on the histopathological changes in the central nervous system (CNS), and this has provided the usual basis for confirmation of the diagnosis of clinical BSE (Wells and Wilesmith 1995; Wells et al. 1989). The histopathological changes are neurodegenerative and are typical of those of other TSEs including scrapie of sheep. Vacuoles and spongiform change are usually conspicuous in the neuropil of grey matter as are single or multiple vacuoles within neuronal perikarya. In BSE, spongiform change is the predominant form of vacuolation. Both neuropil vacuolation and neuronal perikaryonal vacuolation are bilaterally distributed and usually symmetrical with a consistent pattern of severity relative to distribution throughout the brain (Wells and Wilesmith 1995; Wells et al. 1992). Both forms of neuroparenchymal vacuolation reach their greatest intensity in specific anatomic nuclei of the medulla oblongata at the level of the obex. The most consistently affected sites are the solitary tract nucleus, the spinal tract nucleus of the trigeminal nerve and the vestibular complex; the peri-aqueductal grey matter of the midbrain also shows severe and consistent vacuolation (Wells and Wilesmith 1995).

Although the overall severity of changes may differ between animals, the pattern of neuroanatomic nuclei affected by vacuolation is remarkably consistent and the lesions are not affected by either the breed of cow, or route or dose of administration of the agent (Wells and Simmons 1996). This consistency is also temporal throughout the course of the epidemic in Britain (Simmons et al. 1996), all of which suggests that BSE is caused by a single strain of TSE agent. Polymorphisms within the octa-peptide repeat sequence of the PrP gene have been identified in cattle (Goldmann et al. 1991) but there is no evidence that common polymorphic variants at this site alter either susceptibility to disease or the patterns of pathology (Hunter et al. 1994).

The vacuolar changes alone are usually sufficient to establish a diagnosis on a single section of the medulla (Wells et al. 1989). However, as with scrapie of sheep, BSE cases may also occur in which vacuolar brain lesions are minimal or undetectable by light microscopy (Wells et al. 1989, 1994). Also, vacuoles within the perikarya indistinguishable from those of BSE, have been reported in neurones of the red, oculomotor and habenular nuclei as an incidental finding in cattle (Fankhauser et al. 1972; McGill and Wells 1993; Wells, et al. 1989). Thus, histopathological diagnosis of BSE must not rely on the presence of occasional solitary vacuolated neurons and even relatively numerous vacuolated neurons in the red nucleus must be disregarded. However, the presence of spongiform change in specific neuroanatomical locations in BSE is sufficiently distinctive to provide for confident histological diagnosis in the brains of most clinically diseased cattle.

Histopathological changes other than vacuolation are not prominent in the brains of cattle with BSE. Gliosis, mainly astrocytosis, is another feature that usually parallels in magnitude and localization the vacuolar change. Detection of gliosis is enhanced by the use of special stains and following immunohistochemical labelling for glial fibrillary acidic protein. Rarely, amyloid plaques, foci of inflammatory change, neuronophagia and neuronal necrosis may also be seen. 'In situ' end labelling of BSE-affected brains has been attempted to show apoptotic changes, but did not suggest widespread neuronal apoptosis (Theil et al. 1999).

Morphometric studies have shown neuronal loss in some nuclei in which vacuolation was also present (Jeffrey and Halliday 1994), although neuronal loss is not usually sufficiently severe to allow detection by subjective examination, and not all sites showing severe vacuolation have quantifiable deficits in neuronal numbers (Jeffrey et al. 1992). In particular, neuronal loss is not evident at the parasympathetic nucleus of the vagus, despite the fact that this is one of the first and most prominent sites of PrP^d accumulation. The nucleus of the vagus nerve is partly involved in the control of heart rate and ruminal activity, both of which are affected in BSE and in scrapie (Austin and Simmons 1993; Austin et al. 1997). Therefore, in contrast with published comment on the relationship between clinical disease and pathology in Creutzfeldt–Jakob disease (DeArmond et al. 1992; DeArmond 1998), accumulation of PrP^d in the neuroparenchymal tissue of ruminants does not correlate with neuronal loss, and neither the abundance or distribution of PrP^d nor neuronal numbers correlates with neurological and clinical deficits.

Immunohistochemical and ultrastructural examination of sites at which neuronal loss occurs in BSE have failed to reveal any co-existing changes, such as synapse loss (Jeffrey et al. 1992) or axon terminal degeneration, that are usually found in brains of scrapie-affected mice (M. Jeffrey, personal observations). The so-called tubulo-vesicular bodies have been described in most TSEs and although present also in BSE-affected cattle and scrapie-affected sheep, they are extremely rare (M. Jeffrey, personal observations; Ersdal et al. 2003).

2.2
CNS Changes in Scrapie-Affected Sheep

As with BSE, there are no specific gross pathological changes seen in scrapie and the histological lesions are also confined to the CNS (Fraser 1976). The lesions are neurodegenerative, with no specific inflammatory changes and no primary degeneration of the white matter. The most striking change is vacuolation of neurones with characteristic single or multiple vacuoles distending the neuronal perikarya. Vacuolation also affects neuronal processes producing the distinctive appearance of spongiform change in the grey matter neuropil.

Although it is principally the vacuolar changes on which the laboratory diagnosis has been based, some considerations need to be made. Vacuolation of neuronal perikarya is not pathognomonic and it is occasionally found in brains of apparently healthy sheep; in such instances, however, the number of vacuoles is typically much smaller than in cases of clinical scrapie (Zlotnic and Rennie 1957, Zlotnik 1962). On the other hand, incidents of scrapie are recorded in which neuronal vacuolation is virtually undetectable by light microscopy (Fraser 1976; Somerville et al. 1997; Chaplin et al. 1998). This is also the case of experimental infection with the SSBP/1 strain, which commonly results in sparse or extremely limited vacuolation (Begara-McGorum et al. 2002). Therefore, while vacuolation per se may not be sufficient for a diagnosis of scrapie, the absence of significant vacuolar changes in the brain cannot refute a clinical suspect or be taken as evidence of absence of scrapie infection.

Unlike BSE, there is considerable variation in the neuroanatomical distribution of vacuolation—and other changes—in sheep scrapie. This fact, while complicating the histopathological diagnosis may create some advantages. Thus, attempts to use vacuolar lesion profiling to identify sheep scrapie strains, as has been done routinely for many years with murine scrapie strains, have been carried out recently, but have met with only limited success. In sheep scrapie, the vacuolar lesion profile appears to be influenced not only by the agent strain, but also by the host PrP genotype and by other unrelated, some apparently individual, factors (Ligios et al. 2002; Begara-McGorum et al. 2002). The age at onset of clinical disease may also affect the magnitude of the vacuolation, though not its profile (Ligios et al. 2002). It is therefore unlikely that the vacuolation profile will permit accurate strain characterization in individual sheep, though it may help to define disease phenotypes when applied to larger groups or to sheep flocks.

Experimental BSE infection of sheep results in widespread CNS vacuolation which is broadly similar to that found in naturally occurring sheep scrapie. Ovine BSE shows consistent high levels of vacuolar change in the red nucleus, subthalamic nucleus, nucleus ventralis of the thalamus, subiculum and putamen nucleus, while there is an absence of vacuolar pathology in the olivary nuclei, cerebral cortex and hippocampus. However, neither the severity nor the distribution of vacuolar changes are sufficiently distinctive to readily permit differentiation between sheep BSE and natural sheep scrapie (Begara-McGorum et al. 2002; Ligios et al. 2002).

As with cattle BSE, vacuolar changes in sheep scrapie are accompanied by other variable and usually less conspicuous microscopic features, which include other forms of neuronal degeneration, notably the occurrence of dark, shrunken neurones (Hadlow et al. 1982) and also neuronal loss, gliosis (particularly reactive astrocytosis) and amyloidosis. There have been no systematic studies on neuronal loss in sheep scrapie; although some reports have described apparent neuronal loss (Beck et al. 1964), at least one of the sites where this was reported to occur, the cerebellar nodulus, is a brain region with variable neuronal densities in normal sheep. Astrocytosis is also an inconsistent feature of sheep scrapie (Mackenzie 1983). Thus, none of the individual pathological features of scrapie can be considered strictly specific, but in combination and with abundant vacuolar change they are undoubtedly pathognomonic. There is also variation in the prominence of each of the pathological features among individual cases of scrapie, and the appearance of the clinical signs is not necessarily reflected in the severity of the pathology.

3 Immunohistochemistry

For the reasons detailed above, immunohistochemistry (IHC) on tissue sections and/or immunoblotting methods on fresh tissues should be carried out in parallel with routine histology to demonstrate accumulation of abnormal PrP in suspected cases. PrP^{res} can be demonstrated by electrophoresis and immunoblotting in extracts of unfixed brain, previously subjected to the action of detergents and enzymes (Farquhar et al. 1989; Mohri et al. 1992). Preliminary evidence is accumulating to suggest that differences in the relative proportions of the three glycoforms and in the molecular weight of the unglycosylated fragment of PrP^{res} may vary between some sheep TSE sources (Hope et al. 1999; Baron et al. 2000; Stack et al. 2002). These biochemical features, however, are also influenced by host factors (Somerville 1999), and therefore immunochemical methods may not always provide a definitive differentiation between TSE agents and strains.

3.1 Detection of PrP^d in the CNS

When BSE-infected brain tissue of cattle is examined by IHC variations in the cellular patterns of PrP^d accumulation can be found. Intraneuronal, perineuronal, linear, fine puncate and coarse particulate patterns of labelling (Fig. 1) are all present within individual brains (Wells and Wilesmith 1995), but vascular amyloid and classical plaques are extremely rare in cattle brains. PrP^d accumulation generally correlates in distribution with vacuolation, although in individual cow brains the former is usually more widely distributed than vacuolation. As with vacuolation, there are remarkably consistent patterns of PrP^d accumulation in BSE-affected cattle, which is in contrast with the variations observed in sheep scrapie.

The accumulation of PrP^d in scrapie-affected brain is also demonstrated by IHC and can be carried out on routine formalin-fixed material following a variety of protocols, including epitope-demasking techniques and appropriate PrP antibodies (Haritani et al. 1994). The majority of the extracellular PrP^d deposited in brain, including that found in mature amyloid fractions, contains the whole of the N terminus of the protein and is not truncated (Jeffrey et al. 1996). The most intense and

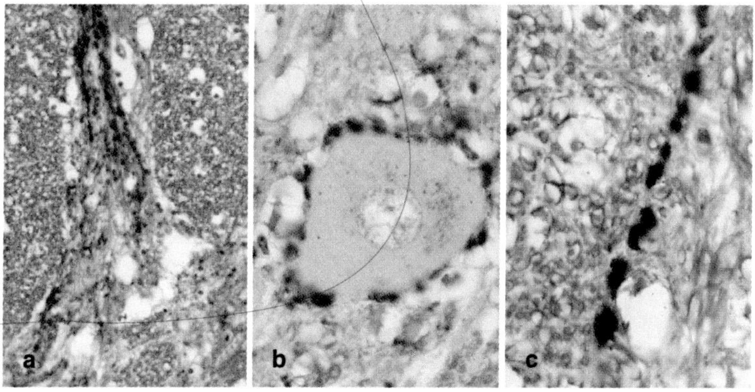

Fig. 1a–c Bovine brain tissue infected with BSE agent. Different patterns of PrP^{d} accumulation are illustrated. **a** Intense localized punctate accumulation in the grey matter neuropil of the spinal tract of the trigeminal nerve. **b** Perineuronal pattern of PrP^{d} accumulation in the reticular formation. **c** Interrupted linear foci of accumulation in the substantia nigra. **a, b** At the level of the obex and **c** from midbrain. IHC for PrP^{d}

consistent accumulations of PrP^{d} are found in the hindbrain and spinal cord, while the variation in intensity of PrP^{d} deposition in forebrain areas is considerable.

Different morphological types of PrP^{d} deposition in the brain of scrapie-affected sheep have been described (van Keulen et al. 1995; Ryder et al. 2001). In recent studies (González et al. 2002, 2003), 13 different morphological types of PrP^{d} accumulation (Fig. 2) were grouped into four different cell-specific, extracellular PrP^{d} patterns and two intracellular, truncated forms of PrP^{d}. The extracellular PrP^{d} patterns were termed neuropil-associated (linear, fine punctate, coarse particulate, coalescing and perineuronal types), astrocyte-associated (stellate, subpial, subependymal and perivascular types), ependymal cell-associated (supraependymal type) and endothelial cell-associated (vascular plaques type). The intracellular patterns were the intraneuronal and intraglial types of PrP^{d} accumulation. Not all these PrP^{d} types and patterns are found in all sheep scrapie cases and the differences in the emerging PrP^{d} profiles can be used in the characterization of TSE agents or strains (see Sect. 5).

Correlation studies of the performance of immunoblot and IHC methods are generally in agreement, but it is clear from screening studies in sheep and cattle that the neuroanatomic sites of examination are critical,

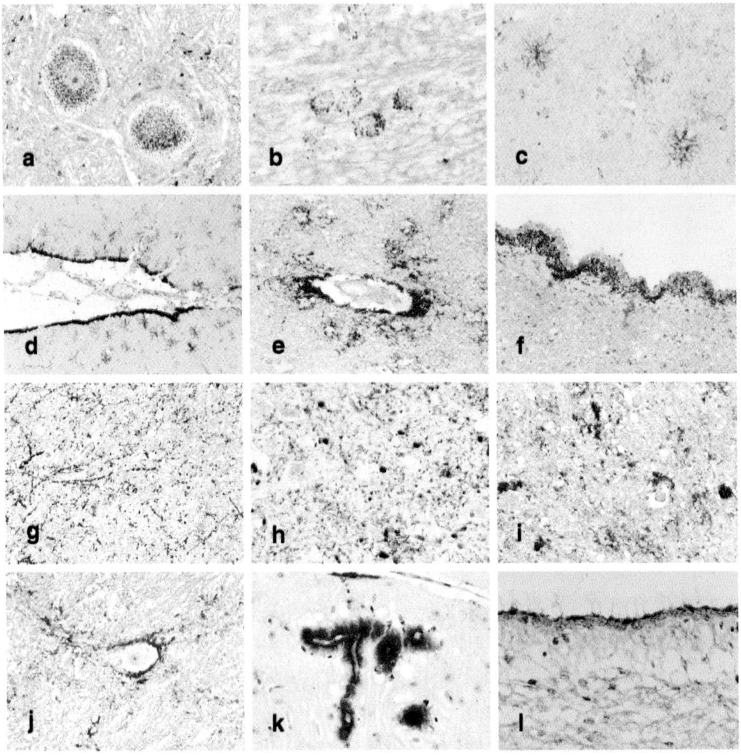

Fig. 2a–l A large range of types of PrP^d accumulation can be found in natural and experimental sheep scrapie. **a** Intraneuronal type: accumulation of granular deposits of PrP^d in the perikarya of neurons of the red nucleus. **b** Intraglial type: accumulation of coarse granular deposits of PrP^d in the cytoplasm of glial cells in the cerebellar white matter. **c** Stellate type: branching deposits of PrP^d on the processes of glial cells in the cerebellar cortex. **d** Subpial type: continuous loose mesh of PrP^d underneath the pia matter in the cerebral cortex. **e** Perivascular type: thick, strongly labelled PrP^d accumulation around a blood vessel in the cerebral white matter. **f** Subependymal type: continuous, strongly labelled mesh of PrP^d underneath the ependyma of the lateral ventricles at the level of the striatum. **g** Linear type: thick thread-like deposits of PrP^d in the neuropil at the level of the obex. **h** Coarse particulate type: irregular, conspicuous deposits of PrP^d in the neuropil at the level of the midbrain. **i** Coalescing type: amorphous, strongly labelled masses of PrP^d in the neuropil at the level of the obex. Note concurrence with coarse particulate deposits. **j** Perineuronal type: thin deposits of PrP^d around the plasmalemma of a neurone in the fastigial nucleus of the cerebellum. **k** Vascular plaques: radiate, fibrillar accumulations of PrP^d around blood vessels in the cerebellar cortex. **l** Ependymal type: continuous, strong PrP^d labelling in the luminal side of the ependymal lining of the lateral ventricle. Note absence of subependymal, astrocyte-associated PrP^d accumulations. (From González et al. 2002 and González, unpublished results). IHC for PrP^d

at least in pre-clinically affected animals (Schaller et al. 1999; Jeffrey et al. 2002). Abnormal PrP can also be demonstrated in unfixed brain extracts in the form of scrapie-associated fibrils visualized by negative stain electron microscopy (Gibson et al. 1987; Scott et al. 1987; Stack et al. 1996). This last technique may be particularly useful when available brain tissue is unsuitable for histological examination due to post mortem autolysis (Stack et al. 1993). Recently modifications of these techniques have also permitted recognition of scrapie-associated fibrils in fixed tissues (Chaplin et al. 1998).

3.2 Detection of PrP^d in Tissues Outwith the CNS

Although cattle BSE and sheep scrapie show many similarities in the nature of tissue changes and PrP^d accumulation in the CNS, there are quite marked differences in the distribution of PrP^d and infectivity in peripheral tissues. Cattle naturally infected with BSE show no infectivity or PrP^d accumulation outwith the CNS. After experimental challenge with high doses of BSE agent, cattle show low levels of infectivity in bone marrow (Wells et al. 1999), peripheral ganglia (Wells et al. 1998) tonsil (Prince et al. 2003) and gut (Wells et al. 1994). In most of these studies, however, the distribution of infectivity has been assessed by cattle-to-mice bioassay, a method that may be 1000 times less sensitive than cattle-to-cattle bioassay (Moynagh and Schimmel 1999). Detection of PrP^d by IHC in experimentally infected cattle seems to be confined to the ileal Peyer's patches (Terry et al. 2002). In contrast with natural sheep scrapie (see below), IHC immunolabelling in this last study was found in only a small proportion of lymphoid follicles where it was confined to macrophage-like cells.

In contrast with BSE of cattle, scrapie-affected sheep show a much wider distribution of tissue infectivity and of PrP^d accumulation. In all but one genotype of sheep affected with scrapie there is widespread PrP^d accumulation in tissues of the lymphoreticular system (LRS), peripheral nervous system (PNS) (van Keulen et al. 1996; Andreoletti et al. 2000; Jeffrey et al. 2001a) and in placenta (Jeffrey et al. 2001a). Natural scrapie does not commonly involve sheep of the VRQ/ARR genotype, but for the small numbers of cases reported, accumulations of PrP^d outwith the CNS are lacking (van Keulen et al. 1996; Andreoletti et al. 2000; Jeffrey et al. 2002). However, some weak peripheral tissue PrP^d accumulation is described following experimental oral challenge of this genotype with the SSBP/1 isolate (Houston et al. 2002). Sheep of the VRQ/ARR genotype therefore appear to have a peripheral pathogenesis more similar to that of BSE in cattle than to that of sheep scrapie in other genotypes. When BSE agent is inoculated into sheep the distribution of peripheral PrP^d is similar to that of natural scrapie, with most sheep showing widespread PrP^d accumulation in many tissues of the LRS (Jeffrey et al. 2001b). As in natural scrapie, a minority of ARQ homozygous sheep infected with BSE do not show any peripheral PrP^d accumulation (Jeffrey et al. 2001b).

All lymphoid tissues are equally affected in terminal disease except for thymus, which does not show significant PrP^d accumulation. Accumulation of PrP^d in the LRS of sheep is mainly restricted to secondary follicles where it is associated with tingible body macrophages (TBMs) and follicular dendritic cells (FDCs) (Jeffrey et al. 2001a). Some macrophage-like cells within adjacent para-cortical T-cell areas (Andreoletti et al. 2000; Heggebø et al. 2002) and in the marginal zone of the spleen (Heggebø et al. 2002) also show PrP^d accumulation. By comparison with studies of scrapie-infected murine LRS (Jeffrey et al. 2000), PrP^d accumulation within the light zone of secondary follicles of sheep is likely to be associated with the plasmalemma or the extracellular space around FDC dendrites, and is labelled by all antibodies tested so far. In contrast, the PrP^d found in the cytoplasm of TBMs is not immunolabelled by peptide specific antibodies raised to the N terminus of PrP, and it is likely therefore, that it represents truncated intra-lysosomal PrP^d (Jeffrey et al. 2001b). Studies using the histoblot technique have confirmed that the PrP^d accumulations in the light zone of secondary follicles and in TBMs of the dark zone contain PrP^{res} (Heggebø et al. 2002).

Accumulation of PrP^d can be found within nervous tissue of the sensory retina (Hardt et al. 2000; Jeffrey et al. 2001a). In the PNS PrP^d can be observed in a variety of sites including major nerve trunks (Groschup et al. 1999), the enteric nervous system (van Keulen et al. 1999), peripheral and cranial nerve ganglia (Jeffrey et al. 2001a) and other sites including the adrenal medulla (Jeffrey et al. 2001a). As in the CNS, PrP^d accumulations found in peripheral ganglia occur both within neuronal perikarya and in association with satellite cells (Jeffrey et al. 2001a).

4 Detection of Infection During the Preclinical Period

Only one study of the temporal evolution of BSE in cattle has so far been completed. In that study, cattle were given a large oral dose of BSE-affected cattle brainstems, but PrP^d detection within the CNS occurred only 3 months before onset of clinical disease at 35 months (Wells et al. 1998). Nevertheless, the patterns and distribution of initial PrP^d accumulation in the CNS are similar in cattle BSE and in sheep and hamster scrapie, which also supports an alimentary route of BSE infection. Mouse bioassay studies complemented with IHC examinations have shown infectivity and PrP^d in the distal ileum of experimentally fed cattle from 6 months after challenge (Wells et al. 1998; Terry et al. 2002). More recently, infectivity has been demonstrated in the tonsil of cattle 10 months after experimental exposure to BSE by the oral route (Prince et al. 2003). Nevertheless, in natural BSE cases and in cattle given low doses of BSE agent, no peripheral PrP^d accumulation or other evidence of infectivity outside the CNS has been found during the preclinical period.

In an environment heavily contaminated with scrapie, sheep of the VRQ/VRQ genotype accumulate PrP^d in LRS tissues (tonsils, gut-associated lymphoid tissue of the ileum and distal jejunal lymph node) as young lambs. It can be found as early as 3 months of age, preceding the earliest evidence of PrP^d in the CNS by several months (Andreoletti et al. 2000; van Keulen et al. 1996). However, in two separate studies of natural scrapie in ARQ/ARQ Suffolk sheep (Jeffrey et al. 2001a) and of experimental BSE in ARQ/ARQ Romney sheep (Jeffrey et al. 2001b), significant PrP^d accumulations were not found until 8 and 10 months, respectively. This may reflect a difference in agent replication rate in ARQ/ARQ genotypes when compared with that of sheep bearing VRQ alleles, though factors such as dose, age at and route of exposure and strain of agent may all contribute to the apparent discrepancies between these studies. The above studies describing the IHC tissue distribution of PrP^d in ARQ/ARQ sheep correlate with earlier mouse bioassay studies of infectivity, originally performed on naturally infected Suffolk sheep (Hadlow et al. 1982). Although this last study was performed prior to PrP genotyping, Suffolk sheep do not have the valine allele at codon 136 and scrapie susceptibility in this breed is almost entirely linked to the ARQ allele (Hunter 1997). Detectable infectivity in these sheep was not found in

7–8-month-old lambs; in 10–14-month-old lambs there was negligible infectivity present in the CNS, while the levels of infectivity in the alimentary tract were comparable to those found in clinically affected animals. These results are in agreement with those of murine scrapie, where in many models infectivity in peripheral tissues reaches a plateau substantially before it does so in the CNS (Kimberlin and Walker 1979). Tissue bioassay of infectivity has also been carried out on BSE-infected ARQ/ARQ sheep with results that indicate a comparable sensitivity of this and the IHC methods. Bioassay studies have not been reported for VRQ sheep, but predictions based on the early onset of PrP^d accumulation in tissues of scrapie-affected sheep with this genotype would suggest that infectivity would occur earlier in the incubation period than in ARQ/ARQ genotypes.

Within the PNS, PrP^d accumulation occurring early in the incubation period may be found in sites which contain mostly sympathetic fibres (adrenal medulla, coeliac-mesenteric ganglia) and also in parasympathetic components of the extrinsic enteric nervous system. In the brain, PrP^d is initially detected in the dorsal motor nucleus of the vagus (DMNV) and in the intermediolateral columns (IMLC) of the spinal cord. These findings taken together suggest that neuroinvasion arises from both parasympathetic (DMNV) and sympathetic (IMLC) peripheral nerves, probably from the alimentary tract (van Keulen et al. 2000; Jeffrey et al. 2001b), which is in agreement with similar more detailed studies of experimental infection in hamsters (McBride et al. 2001; Beekes and McBride 2000).

As described in Sect. 3.2, under natural disease circumstances, sheep of the VRQ/ARR PrP genotype (Andreoletti et al. 2000; van Keulen et al. 1996; Jeffrey et al. 2002) and also a proportion of sheep of the ARQ/ARQ genotype (Jeffrey et al. 2001b) lack detectable infectivity in the peripheral tissues as determined by IHC, bioassay and/or immunoblotting. Similarly, mice with genetic deficiency of lymphoreticular components may also die without PrP^d accumulation or agent replication in the LRS (Fraser et al. 1996; Brown et al. 1996). These results suggest that involvement of the LRS may be influenced by genotype, and that neuroinvasion may occur independently, perhaps even when LRS amplification of infection does take place. However, the tissue compartment and cellular locations at which neuroinvasion may occur are not known. Recently, we have detected the presence of nerve fibres within the secondary follicles of Peyer's Patches, which indicates that infection in lymphoid tissues may facilitate neuroinvasion (Heggebø et al. 2003). Observations in our own laboratory suggest a correlation between PrP^d immunolabelling of ganglia of the enteric nervous system and adjacent PrP^d positive Peyer's patches at early stages of infection. At present, however, there is insufficient data available to determine whether sheep with early LRS involvement have shorter incubation periods than those that do not acquire peripheral LRS infection.

At early stages of infection, PrP^d accumulation within the LRS is most conspicuous within TBMs located in the light zone of the secondary follicles and occurs in the absence of any PrP^d associated with FDCs (Jeffrey et al. 2001a, 2001b). It is likely that small amounts of PrP^d which lie beneath current sensitivity levels of IHC detection are released from infected FDCs and concentrated in TBM lysosomes, making PrP^d detectable. In two separate studies, diffuse PrP patterns of immunolabelling were found within the secondary follicles of gut associated lymphoid tissue (Heggebø et al. 2000), spleen and retropharyngeal lymph node (Jeffrey et al. 2001b). These patterns were found in the absence of any TBM-like PrP^d immunolabelling from weeks to months after experimental challenge with BSE agent (Jeffrey et al. 2001b) or natural scrapie (Heggebø et al. 2000), and were not consistent with morphological patterns of PrP^d accumulation. Although the significance of this immunolabelling is not currently understood, it is likely to represent PrP^c expression on FDCs, which may be increased following exposure to a TSE inoculum containing PrP^d, or in other words, that it represents some up-regulation of PrP^c following exposure to the infectious agent (Heggebø et al. 2000).

Detection of PrP^d by IHC methods is considered to have potential for the pre-clinical in vivo diagnosis of scrapie, using biopsies of tonsil, nictitating membrane or other LRS tissues, such as submandibular lymph node. The detection of PrP^{res} by immunoblotting in samples of placenta (often too autolytic to allow IHC examination) may also offer a non-invasive means of ante mortem diagnosis and disease surveillance in sheep flocks (Race et al. 1998).

5 Strain Diversity in TSEs of Cattle and Sheep

As described above, the uniformity of both vacuolar lesion pattern and PrP^d accumulation in individual animals suggest that cattle BSE is caused by a single strain of TSE agent. Mice inoculated with brain from

UK and Swiss BSE cases showed consistent incubation periods and lesion profiles, which were distinct from those of other TSE sources previously characterized (Bruce et al. 1994, 1996). The biological properties of the BSE agent are retained even after it has been presumptively or experimentally transmitted through intermediate hosts, including man, exotic ungulates, cats and sheep (Bruce et al. 1994). These findings argue that BSE is a novel single strain of TSE agent with high potential for cross-species transmission.

In the USA, experimental inoculations of cattle have been carried out using inocula obtained from sheep scrapie or transmissible mink encephalopathy sources. Tissues obtained from these sources do not readily transmit via the oral route and, although they are able to induce disease when inoculated via parental routes, the lesions developed do not resemble those of cattle BSE (Robinson 1993; Cutlip et al. 1997, 2001). These studies therefore show that, although more than one TSE agent is capable of replicating and causing disease in cattle, only the BSE agent has so far been transmitted to cattle via the oral route and it produces a distinct pathological phenotype. This further reinforces the notion of the novelty and the uniqueness of BSE.

The situation relating TSE strain diversity in sheep is complex. There is clear historical data to support the existence of more than one laboratory or experimental scrapie strain in sheep and recent evidence now suggests that several naturally occurring scrapie strains may also exist. A large number of different murine scrapie strains have been well characterized following transmission from sheep scrapie sources, although it is presently unclear how these strains relate to those causing natural disease of sheep in the field. It remains possible that the isolated murine strains are 'mutations' created on inter-species passage. Previous studies have shown that the experimental transmission of well-characterized pools of brain tissue designated CH1641 and SSBP/1 target different genotypes of sheep or result in very different incubation periods in the same genotype (Hunter 1998). However, the degree to which sheep scrapie strains exist in natural sheep scrapie is not currently well understood. TSE strain recognition in sheep has assumed more importance in recent years with the concern that BSE agent may have entered the sheep population. Although there is no evidence to date from transmission studies of natural sheep TSEs to mice (Bruce et al. 2002) that BSE can be present in sheep, the numbers of experiments completed so far have been too few to draw meaningful data.

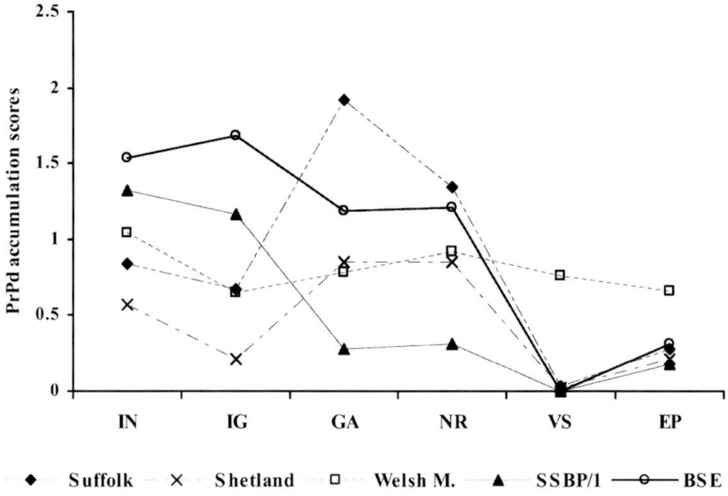

Fig. 3 PrP^d profiles of five groups of clinically affected sheep presumably or actually infected with five different TSE agents or sources. Each point represents the magnitude of accumulation of PrP^d (immunohistochemistry with R486 antibody) in the following cell-associated locations: *IN*, intraneuronal; *IG*, intraglial; *GA*, glia-associated; *NR*, neuropil; *VS*, vascular endothelia; *EP*, ependyma. SSBP/1-infected group includes 31 sheep of different breeds and PrP genotypes, all infected by the subcutaneous route. BSE-infected group includes 24 ARQ/ARQ sheep of a variety of breeds and routes of exposure. Suffolk group includes six ARQ/ARQ sheep, Welsh Mountain group seven VRQ/VRQ sheep and Shetland group nine VRQ/ARQ sheep; all these three groups are naturally infected

The neuroanatomic patterns of PrP^d distribution are one means by which murine scrapie strains can be differentiated. However, differences in the topographic distribution of PrP^d in the brains of outbred sheep populations do not readily allow for strain characterization. As described in Sect. 3.1, there is a large range of morphological types of PrP^d accumulation that can be found in scrapie-affected sheep brains. When these different PrP^d types are grouped into PrP^d patterns according to the cell types involved, the relative magnitudes of each PrP^d pattern provide a PrP^d profile (Fig. 3) (González et al. 2002). The analysis of the PrP^d profiles between groups of sheep exposed to different scrapie sources indicated an effect of the scrapie source on the PrP^d profile (Fig. 3) and a minor effect of the host PrP genotype on the profile and magnitude of PrP^d accumulation, apparently related to the incubation period. Further studies have indicated that the PrP^d profile is not affected by the route of administration and that BSE experimentally affected sheep of the ARQ/ARQ genotype have a PrP^d profile in brain which is different from all scrapie sources examined so far (González et al. 2003 and Fig. 3). Furthermore, ongoing unpublished observations indicate that the PrP^d profile of BSE-affected sheep is very consistent and is not or very little influenced by factors such as breed of sheep, PrP genotype, source and type of inoculum and route of challenge. These results show in summary that differentiation between sheep TSE agents is possible by topographical and cellular analysis of PrP^d deposition in the brain and suggest the occurrence of several natural strains of scrapie agent.

6
Cellular Pathogenesis of Sheep and Cattle TSEs

According to the prion theory, PrP^{sc} is the sole component of the infectious particle (Prusiner 1999b), so that molecular differences in PrP^{sc} would encipher different TSE pathological phenotypes. It has been suggested that variations in molecular weight of PrP^{res} (Bessen and Marsh 1994) and in the degree of asparagine linked glycosylation, as revealed by proteinase K treated immunoblots (Collinge et al. 1996; Telling et al. 1996; Parchi et al. 1998) may form the underlying basis of agent strain differences. Different glycoforms of PrP^{sc} and/or resultant differences in the physiochemical properties of PrP^{sc} molecules would therefore carry the informational component of the infectious agent which is translated into differences in the incubation periods and pathology seen with various scrapie strains or isolates. In turn, these differences would be the consequence of differing rates of production, accumulation and clearance of the prion protein by different cell types within the brain (Safar et al. 1998). It follows that cell specific variation in conformation and/or degree of glycosylation of PrP^{sc} would be the means by which different strains were sustained and propagated (Weissmann 1992; Prusiner 1999b).

Several groups have now reported species, tissue and strain variation in the physio-chemical properties of PrP^c and PrP^{res}. Scrapie mouse brain cells may be infected with different scrapie strains in vitro retaining the properties of each of the original inocula in terms of PrP^{res} glycoform produced and mouse bioassay characteristics (Birkett et al. 2001). These findings show that individual cells are capable of sustaining infection with more than one scrapie strain with subsequent retention of the physiochemical properties and transmission characteristics of the particular strain. Particular molecular signatures of strains, including but not limited to different PrP^{res} glycoforms, are now described in a range of different TSEs (Aguzzi and Weissmann 1997). It has previously been shown that different murine scrapie strains have distinct signatures of PrP^{d} accumulation in the brain (Bruce et al. 1989; Bruce and McBride 1990). These signatures are not route, dose or genotype dependent and it has reasonably been suggested that differential neuronal targeting is the basis for strain variation in mice. On the other hand, differences in glycosylation are found between PrP^{c} obtained from spleen and brain of sheep (Somerville et al. 1997) and three distinct PrP^{c} glycoforms have been found in three discrete neuroanatomical mouse (Somerville 1999) and hamster brain regions (DeArmond et al. 1999). Also, different brain regions of mice infected with the same agent, either scrapie or BSE, show differences in PrP^{res} glycoforms (Somerville 1999) and similar topographical variations have been described in the distribution of PrP^{d} in brains of human TSEs. It has been hypothesized that specific glycoform patterns follow infection of specific subsets of brain nerve cells (DeArmond et al. 1997, 1999). Taking together the results of these studies, it would appear that the physico-chemical, and probably other, properties of PrP^{res}/PrP^{d} depend on the infecting TSE agent and strain and on the cell types that sustain the infection.

6.1
Strain Effects on Targeting

In apparent contrast with murine and human TSE, variation in neuronal tropism does not appear to play a major role in the PrP^{d} accumulation patterns in sheep brains. Rather, the ability of the PrP^{d} profiling methods (described in the preceding section) to discriminate between different sheep TSE sources or strains appears to be based on differences observed in the relative amounts of PrP^{d} associated with cells of different lineages within the brain (González et al. 2002, 2003). Such differences suggest that sheep scrapie and the BSE agent have different affinities for, or ability to target, different cell types. As an example, the agent that causes scrapie in a flock of Suffolk sheep in Scotland infects astrocytes with high avidity, whereas the SSBP/1 scrapie strain shows a high affinity for neurones (Fig. 3). In contrast, when the detailed topography of neuronal populations expressing intracytoplasmic accumulations of PrP^{d} after infection were compared, significant differences were not found between different TSE strains (González et al. 2003). These results indicate that different pathological phenotypes in sheep arise because of variation in the efficiency of infection of different cell types or lineages rather than their propensity to infect different groups of susceptible or unsusceptible neurones. Different glycoforms associated with various sheep TSE infections (Stack et al. 2002) may not be therefore be ascribed to differences in targeting of neurones alone.

6.2 Strain Effects on PrP^d Processing

Different truncated forms of abnormal PrP may occur following infection with different TSE sources as has been suggested from studies of immunoblot patterns of TSEs in man (Parchi et al. 1996) and other experimental models (Bessen and Marsh 1994). Some of the variability in PrP^d profiles observed in the brains of sheep subjected to different TSE infections cannot be explained by differences in cell targeting alone and are also due to variation in truncation of abnormal PrP.

It has been recently found that, for infection with a given TSE source, different peptide-specific antibodies had greater or lesser efficiency in detecting intra-neuronal and intra-microglial accumulation of PrP^d (González et al. 2003). In the same study, however, it was also found that the ability of each of those antibodies to detect intracellular PrP^d deposits also varied when different scrapie sources and BSE were compared (Fig. 4). In other words, each ovine TSE source gave a characteristic signature of intracellular PrP^d accumulation with a panel of PrP antisera (Fig. 5), which is likely to be due to differences in intracellular truncation and digestion of the N terminus of PrP^d after infection with different TSE strains. Differences in PrP^d processing relate not only to intracellular PrP^d truncation, but also to the rate of release of intracellular and/or re-internalization of extracellular PrP^d (González et al. 2003). As an example, in SSBP/1 scrapie infection there is little PrP^d in the neuropil but prominent intra-neuronal PrP^d accumulation, whereas in another two scrapie sources and in sheep BSE very similar levels of these two PrP^d types (intraneuronal and neuropil) are found. These variations in abundance of PrP^d types when different sources are compared suggest different equilibria in the intra-cytoplasmic processing, intracellular trafficking of PrP^d to the cell surface and subsequent release of PrP^d into the neuropil.

The effect of TSE strain on PrP^d processing in sheep is also evident in tissues of the LRS, at least when ovine BSE is compared with a variety of

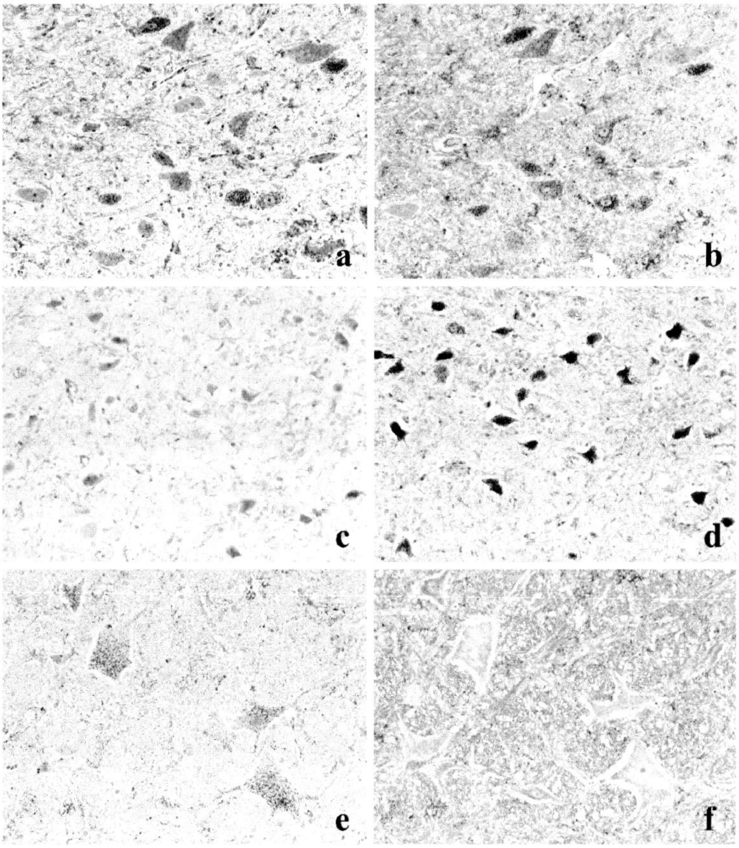

Fig. 4a–f The intensity of intraneuronal PrP^d accumulation is shown in serial or semi-serial sections of brain from three sources of TSE infection in sheep and labelled with two antibodies. Hypoglossal nucleus in natural Welsh Mountain sheep scrapie with 505.2 (**a**) and P4 (**b**); olivary nucleus in SSBP/1 scrapie with 505.2 (**c**) and P4 (**d**); reticular formation in experimental sheep BSE with 505.2 (**e**) and P4 (**f**). There is similar immunoreactivity with the two antibodies in Welsh Mountain sheep, lack of, or strong reactivity to 505.2 and P4, respectively in SSBP/1-affected sheep but the opposite results are present in BSE-affected sheep. Note also absence of PrP^d accumulation in the neuropil of SSBP/1-affected sheep in spite of high levels of intracellular PrP^d accumulation. IHC for PrP^d, ×120

natural and experimental scrapie strains infections (Jeffrey et al. 2001, 2003). Phagocytic cells within the LRS of scrapie-affected sheep show no immunolabelling with antibodies raised against extreme downstream segments of the N terminus of the PrP molecule, but show consistent la-

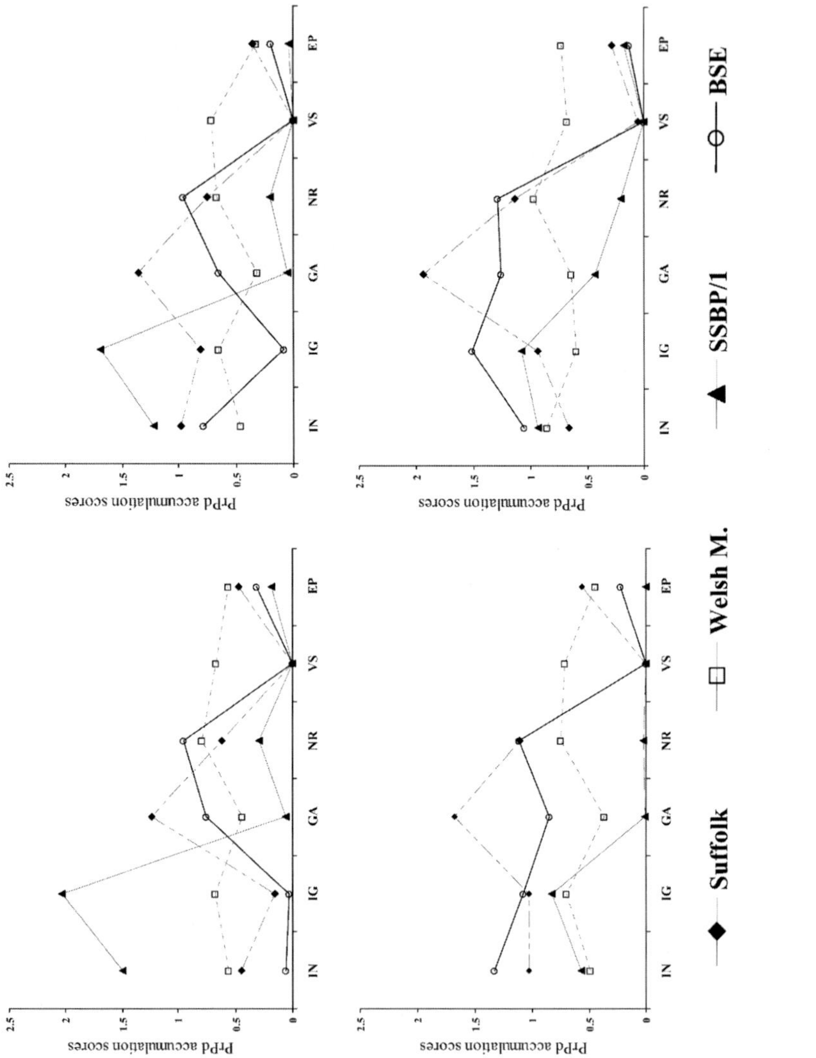

Fig. 5 PrP^d profiles of the four sheep groups when labelled with four different antibodies. The graphs illustrate variation in the antibody binding affinity for different cell-associated patterns of PrP^d accumulation. Different affinities suggest that different processing and conformation of PrP^d occur in neurons as shown in Fig. 4 and also in non-neuronal cell types; compare glia-associated binding affinity in Suffolk sheep, SSBP/1- and BSE-infected sheep. *IN*, intraneuronal PrP^d; *IG*, intraglial PrP^d; *GA*, glia-associated PrP^d; *NR*, neuropil PrP^d; *VS*, vascular PrP^d; *EP*, ependymal PrP^d. SSBP/1-infected group includes five VRQ/VRQ Cheviot sheep infected by the subcutaneous route. BSE-infected group includes five ARQ/ARQ Romney sheep infected intracerebrally. Suffolk group includes five ARQ/ARQ sheep and Welsh Mountain group five VRQ/VRQ sheep; these two last groups are naturally infected

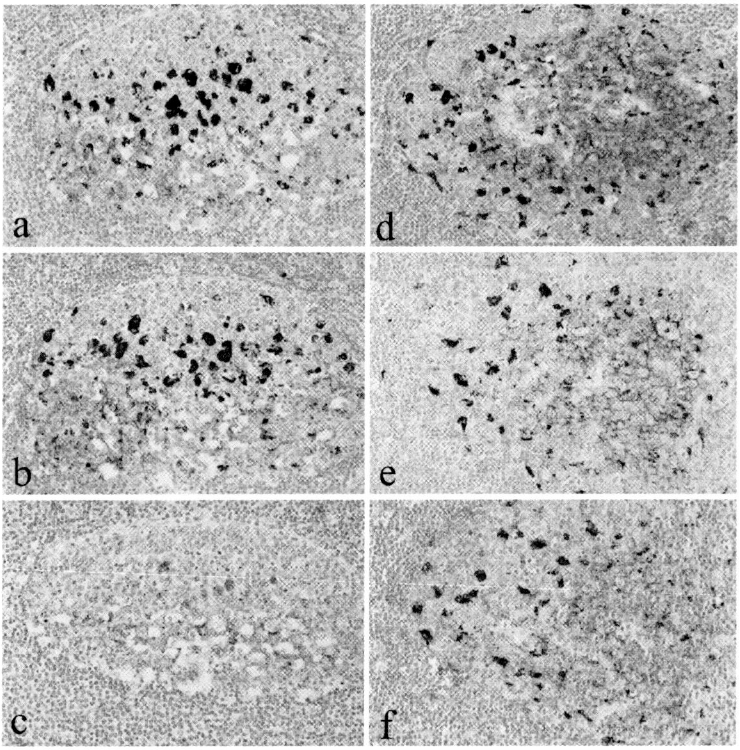

Fig. 6a–f A comparison of retropharyngeal lymph node taken from sheep infected with either scrapie or BSE. Serial sections are of the same secondary follicle taken from a BSE agent infected ARQ/ARQ Romney sheep (**a–c**) and similarly, serial sections from the same secondary follicle of a scrapie-infected Suffolk sheep (**d–f**) are shown. The sections are labelled with antibodies R486 (**a**, **d**), L42 (**b**, **e**) and 521.7 (**c**, **f**). In the Suffolk sheep there is labelling of the light zones with marked-FDC type labelling and also an intense granular labelling associated with individual cells (TBM) located in the light zone, dark zone and follicular mantle. In the follicle from the BSE agent infected sheep there is a similar pattern with the R486 (**a**) and L42 (**b**) but with the 521.7 antibody there is almost exclusively a weak FDC type pattern with little or no evidence of TBM labelling (**c**). ×120

belling with antibodies to the upstream segments of the flexible tail of the same molecule. In contrast, phagocytic cells of BSE-infected sheep tissues lack labelling with antibodies to both downstream and upstream segments of the flexible tail of the PrP molecule, the differences with scrapie being particularly evident in the 84–111 amino acid sequence (Fig. 6) (Jeffrey et al. 2001). This, therefore, indicates that strain-dependent processing of PrP^{d} in specific cell types within the LRS can be used to distinguish between ovine BSE and scrapie-infected sheep. As immunolabelling of LRS phagocytic cells is found in the lysosomes, absence of immunolabelling at such sites with antibodies recognizing downstream segments of the flexible tail of the PrP molecule suggests differences in the truncation sites between BSE and scrapie PrP^d molecules. Such differences in truncation must reflect strain specific variation in the conformation of the PrP^d molecule at its source, which in the case of the intracellular PrP^d found in TBMs are FDCs (Jeffrey et al. 2001, 2002).

6.3
Cell Effects on PrP^d Truncation

PrP^d peptide mapping studies suggest that variation in the truncation site of PrP^d in TSE-infected sheep may not be entirely explained by TSE agent or strain diversity. As described above, the intracellular labelling affinity of TBMs in BSE-infected sheep LRS tissues differs from that of scrapie-affected sheep. However, when different phagocytic cells and neurones of ovine BSE-infected tissues are examined, the precise labelling pattern changes with the cell types in different tissue compartments (Jeffrey et al. 2002b). This indicates that the cells internalizing PrP^d for phagocytosis (macrophages, glia and neurones; Fig. 7) may initially truncate PrP^d at different sites within the putative flexible tail region even when the internalized or accumulated PrP^d originates from infection with a single agent. At least three different truncation lengths of PrP^d were inferred from IHC patterns following infection with the BSE agent strain in sheep (Jeffrey et al. 2001, 2002b) suggesting that the processing of PrP^d or the conformation PrP^d is partly dependent on the cell type infected.

Cell biology studies show that PrP, in common with other glyco-phosphatidil-inositol-anchored cell surface proteins, is transported to the lumen of the Golgi apparatus from where, under the influence of chaperones, it is folded and glycosylated within the endoplasmic reticulum before being transported to the cell surface (Harris 1997). Abnormal or misfolded proteins are transported to the endosomal system under the influence of ubiquitin or other heat shock proteins. Different conformers of PrP^d might therefore be detected within the cell cytoplasm using different antibodies after they are transported to the endosomal/lysosomal system or following re-internalization of abnormal PrP^d from the cell surface. In at least the experimental SSBP/1 scrapie strain there is con-

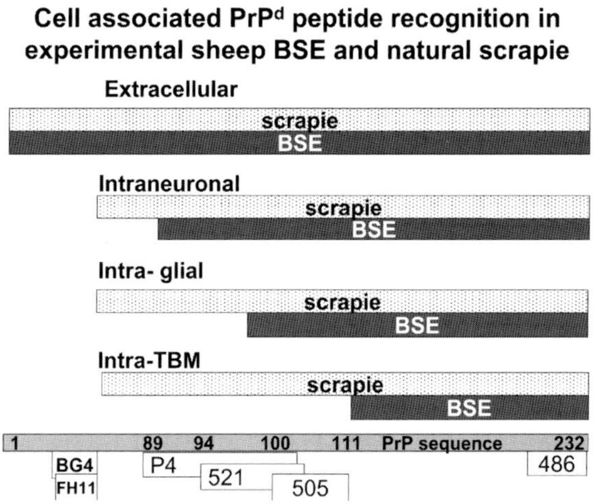

Fig. 7 Schematic diagram summarizing variation in the antibody binding affinity for different sheep BSE sources and sheep scrapie sources for all forms of extracellular PrP^d accumulation, and intracellular PrP accumulation in neurones, macrophages and glia. The diagram shows that there is variation in the labelling of the N terminus for ovine BSE PrP^d in different cell types located in both the same tissue (CNS) and in different tissues. The PrP molecule is represented as a linear sequence of amino acids (*pale grey bar*) relative to the sequence of amino acids used to generate anti-peptide antiserum (*white rectangles*). The *stippled bars* indicate the inferred length of the PrP molecule as shown by anti-peptide antibody immunolabelling to PrP^d accumulations at different cellular sites for scrapie-affected sheep. The *dark grey bars* show the antibody labelling affinity of PrP^d accumulation for BSE-affected sheep

spicuous intra-neuronal PrP^d in the absence of extracellular forms suggesting that the PrP^d may never reach the cell surface. Whatever the precise mechanism, the studies described above suggest that infection of sheep with different TSE agents or strains result in different PrP^d conformers being produced by individual cells and that variation in the processing of PrP^d occurs in different cells following infection with the same TSE source, at least in sheep BSE.

7
Conclusions

Scrapie and BSE are related TSEs or prion diseases with a different history. Whereas sheep scrapie has been recognized in Europe since the eighteenth century, BSE was first recognized in Great Britain in 1986. In

BSE, a highly repeatable topographic pattern of vacuolation and PrP^d accumulation is consistent with disease caused by a single agent strain. In contrast, sheep scrapie shows a highly variable lesion distribution and a diverse profile of PrP^d deposition. Abnormal accumulations of PrP^d generally correlate with infectivity in a range of tissues of scrapie and BSE-affected cattle and sheep. BSE-affected cattle and certain sheep genotypes affected with scrapie show only a restricted range of extra-neural PrP^d accumulation or infectivity, while most scrapie-affected sheep have PrP^d and infectivity throughout the CNS, LRS and PNS. Both sheep and cattle are thought to acquire infection via the oral route, and these marked differences in tissue distribution of infectivity and PrP^d accumulation indicate marked differences in the peripheral pathogenesis of the corresponding diseases. The peripheral pathogenesis of experimental ovine BSE is, however, closely similar to that of natural sheep scrapie.

Different natural and experimental scrapie sources result in different proportions of PrP^d accumulation associated with different cell types within the brain (the PrP^d profile), which are not clearly influenced by sheep genotype and route of inoculation. Different sources of sheep scrapie therefore appear to have different tropisms for different cell types within the CNS. Ovine BSE can be distinguished from other UK sheep scrapie sources by its PrP^d profile and by the absence of labelling of upstream N-terminal segments of PrP^d in neurones and phagocytic cells of the brain and LRS. These differences infer different truncated forms of PrP^d being associated with different ovine TSE sources.

References

Aguzzi A, Weissmann C (1997) Prion Research—The Next Frontiers. Nature 389:795–798

Andreoletti O, Berthon P, Marc D, Sarradin P, Grosclaude J, vanKeulen L, Schelcher F, Elsen JM and Lantier F (2000) Early accumulation of PrP^{Sc} in gut-associated lymphoid and nervous tissues of susceptible sheep from a Romanov flock with natural scrapie. J Gen.Virol 81:3115–3126

Austin AR, Pawson L, Meek S and Webster S (1997) Abnormalities of Heart Rate and Rhythm in Bovine Spongiform Encephalopathy. Vet Rec 141:352–357

Austin AR, Simmons MM (1993) Reduced rumination in bovine spongiform encephalopathy and scrapie. Vet Rec 132:324–325

Baron TGM, Madec JY, Calavas D, Richard Y, and Barillet F (2000) Comparison of French natural scrapie isolates with bovine spongiform encephalopathy and experimental scrapie infected sheep. Neurosci Lett 284:175–178

Beck E, Daniel PM, Parry HB (1964) Degeneration of the cerebellar and hypothalamic neurohypophysial systems in sheep with scrapie; and its relationship to human system degenerations. Brain 87:153–176

Beekes M and McBride PA (2000) Early accumulation of pathological PrP in the enteric nervous system and gut-associated lymphoid tissue of hamsters orally infected with scrapie. Neurosci Lett 278:181–184.

Begara-McGorum I, González L, Simmons M, Hunter N, Houston F, and Jeffrey M (2002) Vacuolar lesion profile is sheep scrapie. Analysis of factors involved in its variation and relationship to disease specific accumulation. J Comp Pathol 127: 59–68

Bessen RA and Marsh RF (1994) Distinct PrP Properties Suggest the Molecular Basis of Strain Variation in Transmissible Mink Encephalopathy. Virol 68:7859–7868

Birkett CR, Hennion RM, Bembridge DA, Clarke MC, Chree A, Bruce M E, and Bostock CJ (2001) Scrapie strains maintain biological phenotypes on propagation in a cell line in culture. EMBO J 20:3351–3358

Bossers A, Belt PBGM, Raymond GJ, Caughey B, deVries R, Smits MA (1997) Scrapie susceptibility-linked polymorphisms modulate the in vitro conversion of sheep prion protein to protease-resistant forms. PNAS 94:4931–4936

Brown KL, Stewart K, Bruce ME, Fraser H (1996) Scrapie in immunodeficient mice. In: Court L, Dodet B (eds) Transmissible Subacute Spongiform Encephalopathies: Prion Diseases. Elsevier Editions Scientifiques, Paris. pp 159–166

Bruce ME and McBride PA (1990) Neuroanatomical distribution of PrP accumulation in mice infected with different strains of scrapie. Neuropath Appl Neurobiol 16:541

Bruce M, Chree A, McConnell I, Brown K and Fraser H (1996) Transmission and strain typing studies of scrapie and bovine spongiform encephalopathy. In: Court L, Dodet B (eds) Transmissible Subacute Spongiform Encephalopathies: Prion Diseases. Elsevier Editions Scientifiques, Paris. pp 259

Bruce M, Chree A, McConnell I, Foster J, Pearson G and Fraser H (1994) Transmission of bovine spongiform encephalopathy and scrapie to mice: strain variation and the species barrier. Phil Trans Roy Soc 343:405–411

Bruce ME, Boyle A, Cousens S, McConnell I, Foster J, Goldmann W and Fraser H (2002) Strain characterization of natural sheep scrapie and comparison with BSE. J Gen Virol 83:695–704

Bruce ME, Fraser H, McBride PA, Scott JR and Dickinson AG (1992) The basis of strain variation in scrapie Prion diseases of humans and animals [edited by Prusiner SB, Collinge J, Powell J, Anderton B.]Ellis Horwood, London

Bruce ME, McBride PA, Farquhar CF (1989) Precise Targeting of the Pathology of the Sialoglycoprotein, PrP, and Vacuolar Degeneration in Mouse Scrapie. Neurosci Lett 102:1–6

Chaplin MJ, Aldrich AD, Stack MJ (1998) Scrapie associated fibril detection from formaldehyde fixed brain tissue in natural cases of ovine scrapie. Res Vet Sci 64:41–44

Collinge J, Sidle KCL, Meads J, Ironside J and Hill AF (1996) Molecular analysis of prion strain variation and the aetiology of 'new variant' CJD. Nature 383:685–690

Cunningham A, Wells G, Scott A, Kirkwood J and Barnett J (1993) Transmissible spongiform encephalopathy in greater kudu (Tragelaphus strepsiceros) Vet Rec 132:68

Cutlip RC, Miller JM, Hamir AN, Peters J, Robinson MM, Jenny AL, Lehmkuhl HD, Taylor WD and Bisplinghoff FD (2001) Resistance of cattle to scrapie by the oral route. Can J Vet Res 65:131–132

Cutlip RC, Miller JM, Lehmkuhl HD (1997) Second passage of a US scrapie agent in cattle. J Comp Pathol 117:271–275

DeArmond SJ, Kristensson K, Bowler RP (1992) PrPsc Causes Nerve Cell Death and Stimulates Astrocyte Proliferation: A Paradox. Prog Brain Res 94:437–446

DeArmond SJ (1998) Prion Diseases—The Spectrum of Etiologic and Pathogenic Mechanisms. In: Folstein MF (ed) Neurobiology of Primary Dementia. American Psychiatric Press, Inc, 1400 K St NW/Washington/DC 20005, pp 83–118

DeArmond SJ, Qiu Y, Sanchez H, Spilman PR, NinchakCasey A, Alonso D and Daggett V (1999) PrPC glycoform heterogeneity as a function of brain region: Implications for selective targeting of neurons by prion strains. J Neuropath Exp Neurol 58:1000–1009

DeArmond SJ, Sanchez H, Yehiely F, Qiu Y, NinchakCasey A, Daggett V, Camerino AP, Cayetano J, Rogers M, Groth D, Torchia M, Tremblay P, Scott MR, Cohen FE and Prusiner SB (1997) Selective neuronal targeting in prion disease. Neuron 19:1337–1348

Ersdal C, Simmons MM, Goodsir C, Martin S and Jeffrey M (2003) Sub-cellular pathology of scrapie: coated pits are increased in PrP codon 136 alanine homozygous scrapie-affected sheep. Acta Neuropathol 106:17–28

Fankhauser R, Fatzer R, Frauchiger E (1972) Spastic paralysis in cattle. Schweiz Arch Tierheilkd 114:24–32

Farquhar CF, Somerville RA, Ritchie LA (1989) Post-mortem immunodiagnosis of scrapie and bovine spongiform encephalopathy. J Virol Met 24:215–222

Farquhar CF, Somerville RA, Bruce ME (1998) Straining the prion hypothesis. Nature 391:345–346

Fraser H (1976) The pathology of natural and experimental scrapie. In: Kimberlin RH (ed) Slow virus diseases of animals and man. North-Holland Publishing Company, Amsterdam, pp 267–305

Fraser H, Brown KL, Stewart K, McConnell I, McBride P, and Williams A (1996) Replication of scrapie in spleens of SCID mice follows reconstitution with wild-type mouse bone marrow. J Gen Virol 77:1935–1940

Gibson PH, Somerville RA, Fraser H, Foster JD, and Kimberlin RH (1987) Scrapie associated fibrils in the diagnosis of scrapie in sheep. Vet Rec 120:125–127

Goldmann W, Hunter N, Martin T, Dawson M and Hope J (1991) Different Forms of the Bovine PrP Gene Have Five or Six Copies of a Short, G-C-Rich Element Within the Protein-Coding Exon. J Gen Virol 72:201–204

González L, Martin S, Begara-McGorum I, Hunter N, Houston F, Simmons M, and Jeffrey M (2002) Effects of agent strain and host genotype on PrP accumulation in the brain of sheep naturally and experimentally affected with scrapie. J Comp Pathol 126:17–29

González L, Martin S, Jeffrey M (2003) Distinct profiles of PrP^d immunoreactivity in the brain of scrapie and BSE infected sheep: implications on differential cell targeting and PrP processing . J Gen Virol 84:1339–1350

Groschup M H, Beekes M, McBride PA, Hardt M, Hainfellner J A and Budka H. (1999) Deposition of disease-associated prion protein involves the peripheral nervous system in experimental scrapie. Acta Neuropathol. 98:453–457

Hadlow WJ, Kennedy RC, Race RE (1982) Natural infection of Suffolk sheep with scrapie virus. Journal of Infectious diseases 146:657–664

Hardt M, Baron T, Groschup, MH (2000) A comparative study of immunohistochemical methods for detecting abnormal prion protein with monoclonal and polyclonal antibodies. J Comp Pathol 122:43–53

Haritani M, Spencer YI, Wells GAH (1994) Hydrated autoclave pretreatment enhancement of prion protein immunoreactivity in formalin fixed bovine spongiform encephalopathy affected brain. Acta Neuropathol 87: 86–90

Harris DA (1997) Cell biological studies of the prion protein. In: Harris DA (ed) Prions: Molecular and Cellular Biology. Horizon Scientific Press, Wymondham, UK, pp 53–65

Heggebø R, González L, Press CM, Gunnes G, Espenes A, Jeffrey M (2003) Disease-associated PrP in the enteric nervous system of scrapie-affected Suffolk sheep. J Gen Virol 84:1327–1338

Heggebø R, Press CM, Gunnes G, González L and Jeffrey M (2002) Distribution and accumulation of PrP in gut-associated and peripheral lymphoid tissue of scrapie-affected Suffolk sheep. J Gen Virol 83:479–489

Heggebø R, Press CM, Gunnes G, Lie KI, Tranulis MA, Ulvund M, Groschup MH, and Landsverk T (2000) Distribution of prion protein in the ileal Peyer's patch of scrapie-free lambs and lambs naturally and experimentally exposed to the scrapie agent. J Gen Virol 81:2327–2337

Hope J, Wood SCER, Birkett CR, Chong A, Bruce ME, Cairns D, Goldmann W, Hunter N and Bostock CJ (1999) Molecular analysis of ovine prion protein identifies similarities between BSE and an experimental isolate of natural scrapie, CH1641. J Gen Virol 80:1–4

Houston EF, Halliday SI, Jeffrey M, Goldmann W and Hunter N (2002) New Zealand sheep with scrapie-susceptible PrP genotypes succumb to experimental challenge with a sheep-passaged scrapie isolate (SSBP/1) J Gen Virol 83:1247–1250

Hunter N (1997) PrP genetics in sheep and the implications for scrapie and BSE. TIMS 5:331–334

Hunter N (1998) Scrapie. Mol Biotech 9:225–234

Hunter N, Goldmann W, Smith G and Hope J (1994) Frequencies of PRP Gene Variants in Healthy Cattle and Cattle with BSE in Scotland. Vet Rec 135:400–403

Jeffrey M and Halliday W (1994) Numbers of neurons in vacuolated and non vacuolated neuroanatomical nuclei in bovine spongiform encephalopathy affected brains. J Comp Pathol 110:287–293

Jeffrey M, Halliday W, Goodsir C (1992) A morphometric and immunohistochemical study of the vestibular nuclear complex in bovine spongiform encephalopathy. Acta Neuropathol 84:651–657

Jeffrey M, Begara-McGorum I, Clark S, Martin S, Clark J, Chaplin M and González L (2002) Occurrence and Distribution of Infection-specific PrP in Tissues of Clinical Scrapie Cases and Cull Sheep from Scrapie-affected Farms in Shetland. J Comp Pathol 127:264–273

Jeffrey M, Martin S, Thomson JR, Dingwall WS, BegaraMcGorum I and González L (2001a) Onset and distribution of tissue PrP accumulation in scrapie-affected suffolk sheep as demonstrated by sequential necropsies and tonsillar biopsies. J Comp Pathol 125:48–57

Jeffrey M, McGovern G, Goodsir CM, Brown KL and Bruce ME (2000) Sites of prion protein accumulation in scrapie-infected mouse spleen revealed by immuno-electron microscopy. J Pathol 191:323–332

Jeffrey M, Ryder S, Martin S, Hawkins SAC, Terry L, Berthelin Baker C and Bellworthy SJ (2001b) Oral inoculation of sheep with the agent of bovine spongiform encephalopathy (BSE) 1. Onset and distribution of disease-specific PrP accumulation in brain and viscera. J Comp Pathol 124:280–289

Jeffrey M and Wells GA (1988) Spongiform encephalopathy in a nyala (Tragelaphus angasi) Vet Pathol 25:398–399

Jeffrey M, Martin S, González L (2003) Cell-associated variants of disease-specific PrP immunolabelling are found in different sources of sheep transmissible spongiform encephalopathy. J Gen Virol 84:1033–1046

Kimberlin RF and Walker CA (1979) Pathogenesis of mouse scrapie: effect of route of inoculation on infectivity titres and dose response curves. J Comp Pathol 89:39–47

Ligios C, Jeffrey M, Ryder SJ, Bellworthy SJ and Simmons M (2002) Distinction of scrapie phenotypes in sheep by lesion profiling. J Comp Pathol 127:45–57

Mackenzie A (1983) Immunohistochemical demonstration of glial fibrillary acidic protein in scrapie. J Comp Pathol 93:251–259

McBride PA, Schulz-Schaeffer WJ, Donaldson M, Bruce M, Diringer H, Kretzschmar HA and Beekes M (2001) Early spread of scrapie from the gastrointestinal tract to the central nervous system involves autonomic fibers of the splanchnic and vagus nerves. J Virol 75:9320–9327

McGill I and Wells GAH (1993) Neuropathological findings in cattle with clinically suspect but histologically unconfirmed bovine spongiform encephalopathy (BSE)J Comp Pathol 108:241–260

Mohri S, Farquhar CF, Somerville RA, Jeffrey M, Foster J and Hope J (1992) Immunodetection of a disease specific PrP fraction in scrapie-affected sheep and BSE-affected cattle. Vet Rec 131:537–539

Moynagh J and Schimmel H (1999) Tests for BSE evaluated. Nature 400:105

Nathanson N, Wilesmith J, Wells GA and Griot C (1999) Bovine spongiform encephalopathy and related diseases. In: Prusiner SB (ed) Prion Biology and Diseases. Cold Spring Harbor Laboratory Press, Plainview, NY, pp 431–463

Parchi P, Castellani R, Capellari S, Ghetti B, Young K, Chen SG, Farlow M, Dickson DW, Sima AAF, Trojanowski JQ, Petersen RB and Gambetti P (1996) Molecular basis of phenotypic variability in sporadic Creutzfeldt-Jakob disease. Ann Neurol 39:767–778

Parchi P, Chen SG, Brown P, Zou WQ, Capellari S, Budka H, Hainfellner J, Reyes PF, Golden GT, Hauw JJ, Gajdusek DC and Gambetti P (1998) Different patterns of truncated prion protein fragments correlate with distinct phenotypes in P102L Gerstmann-Straussler-Scheinker disease. PNAS 95:8322–8327

Prince MJ, Bailey JA, Barrowman PR, Bishop KJ, Campbell GR, and Wood JM (2003) Bovine Spongiform encephalopathy. Rev sci tech Off int Epiz 22:37–55

Prusiner SB (1982) Novel proteinaceous infectious particles cause scrapie. Science 216:136–144

Prusiner SB (1999a) An introduction to prion biology and diseases. In: Prusiner SB (ed) Prion Biology and Diseases. Cold Spring Harbor Laboratory Press, Plainview, NY, pp 1–66

Prusiner SB (1999b) Development of the prion concept. In: Prusiner SB (ed) Prion Biology and Diseases. Cold Spring Harbor Laboratory Press, Plainview, NY, pp 67–112

Race R, Jenny A, Sutton D (1998) Scrapie Infectivity and Proteinase K-Resistant Prion Protein in Sheep Placenta, Brain, Spleen, and Lymph-Node: Implications for Transmission and Antemortem Diagnosis. J Infect Dis 178:949–953

Robinson M (1993) Transmission studies with transmissible mink encephalopathy and bovine spongiform encephalopathy, and a survey of mink feeding practices. J Am Vet Med Assoc 204:72

Ryder S, Spencer YI, Bellerby PJ and March SA (2001) Immunohistochemical detection of PrP in the medulla oblongata of sheep:the spectrum of staining in normal and scrapie-affected sheep. Vet Rec 148:7–13

Safar J, Wille H, Itrri V, Groth D, Serban H, Torchia M, Cohen FE and Prusiner SB (1998) Eight prion strains have PrP^{Sc} molecules with different conformations. Nature Med 4:1157–1165

Schaller O, Fatzer R, Stack M, Clark J, Cooley W, Biffiger K, Egli S, Doherr M, Vandevelde M, Heim D, Oesch B and Moser M (1999) Validation of a Western immunoblotting procedure for bovine PrPSc detection and its use as a rapid surveillance method for the diagnosis of bovine spongiform encephalopathy (BSE) Acta Neuropathol 98: 437–443.

Scott AC, Done SH, Venables C and Dawson M (1987) Detection of scrapie associated fibrils as an aid to the diagnosis of natural sheep scrapie. Vet Rec 120:280–281

Simmons MM, Harris P, Jeffrey M, Meek SC, Blamire IWH and Wells GAH (1996) BSE in Great Britain: Consistency of the neurohistopathological findings in two random annual samples of clinically suspect cases. Vet Rec 138:175–177

Somerville RA (1999) Host and transmissible spongiform encephalopathy agent strain control glycosylation of PrP. J Gen Virol 80:1865–1872

Somerville RA, Birkett CR, Farquhar CF, Hunter N, Goldmann W, Dornan J, Grover D, Hennion RM, Percy C, Foster J and Jeffrey M (1997) Immunodetection of PrPSc in spleens of some scrapie-infected sheep but not BSE-infected cows. J Gen Virol 78:2389–2396

Stack MJ, Scott AC, Done SH and Dawson M (1993) Scrapie associated fibril detection on decomposed and fixed ovine brain material. Res Vet Sci 55:173–178

Stack MJ, Keyes P, Scott AC (1996) The diagnosis of bovine spongiform encephalopathy and scrapie by the detection of fibrils and the abnormal protein isoform. In: Baker HF, Ridley RM (eds) Prion Diseases. Humana Press Inc, Totowa, NJ, pp 85–103

Stack M, Chaplin M, Clark J (2002) Differentiation of prion protein glycoforms from naturally occurring sheep scrapie, sheep-passaged scrapie strains (CH1641 and SSBP1), bovine spongiform encephalopathy (BSE) cases and Romney and Cheviot sheep experimentally inoculated with BSE using two monoclonal antibodies. Acta Neuropathol 104:279–286

Telling GC, Haga T, Torchia M, Tremblay P, DeArmond SJ and Prusiner SB (1996) Interactions between wild-type and mutant prion proteins modulate neurodegeneration transgenic mice. Genes and Dev 10:1736–1750

Terry L, Marsh S, Ryder SJ, Hawkins SAC, Wells GAH and Spencer YI (2003) Detection of disease-specific PrP in the distal ileum of cattle orally exposed to the BSE agent. Vet Rec 152:387–392

Theil D, Fatzer R, Meyer R, Schobesberger M, Zurbriggen A, Vandevelde M (1999) Nuclear DNA fragmentation and immune reactivity in bovine spongiform encephalopathy. J Comp Pathol 121:357–367

Vankeulen LJM, Schreuder BEC, Meloen RH, Mooijharkes G, Vromans MEW and Langeveld JPM (1996) Immunohistochemical detection of prion protein in lymphoid tissues of sheep with natural scrapie. J Clin Micro 34:1228–1231

Vankeulen LJM, Schreuder BEC, Meloen RH, Poelenvandenberg M, Mooijharkes G, Vromans MEW and Langeveld JPM (1995) Immunohistochemical Detection and Localization of Prion Protein in Brain-Tissue of Sheep with Natural Scrapie. Vet Pathol 32:299–308

Vankeulen LJM, Schreuder BEC, Vromans MEW, Langeveld JPM and Smits MA (1999) Scrapie-associated prion protein in the gastrointestinal tract of sheep with natural scrapie. J Comp Pathol 121:55–63

Vankeulen LJM, Schreuder BEC, Vromans MEW, Langeveld JPM and Smits MA (2000) Pathogenesis of natural scrapie in sheep. Archives of Virology suppl 16:57–71

Weissmann C (1992) How can the 'protein only' hypothesis of prion propagation be reconciled with the existence of multiple prion strain? Prion diseases of humans and animals (edited by Prusiner, S B ; Collinge, J ; Powell, J ; Anderton, B) Ellis Horwood Limited, Chichester, UK, 523–530

Wells GAH, Dawson M, Hawkins SAC, Green RB, Dexter I, Francis ME, Simmons MM, Austin AR and Horigan MW (1994) Infectivity in the ileum of cattle challenged orally with bovine spongiform encephalopathy. Vet Rec 135:40–41

Wells GAH, Hancock RD, Cooley WA, Richards MS, Higgins RJ and David GP (1989) Bovine Spongiform Encephalopathy: Diagnostic Significance of Vacuolar Changes in Selected Nuclei of the Medulla Oblongata. Vet Rec125:521–524

Wells GAH, Hawkins SAC, Green RB, Austin AR, Dexter I, Spencer YI, Chaplin MJ, Stack MJ and Dawson M (1998) Preliminary Observations on the Pathogenesis of Experimental Bovine Spongiform Encephalopathy (BSE): An Update. Vet Rec 142:103–106

Wells GAH, Hawkins SAC, Green RB, Spencer YI, Dexter I and Dawson M (1999) Limited detection of sternal bone marrow infectivity in the clinical phase of experimental bovine spongiform encephalopathy (BSE) Vet Rec 144:292–294

Wells GAH, Hawkins SAC, Hadlow WJ, Spencer YI (1992) The discovery of bovine spongiform encephalopathy and observations on the vacuolar changes Prion diseases of humans and animals (edited by Prusiner SB, Collinge J, Powell J, Anderton B)Ellis Horwood, London

Wells GAH and McGill IS (1992) Recently described scrapie-like encephalopathies of animals: case definitions. Res Vet Sci 53:1–10

Wells GAH, Scott AC, Johnson CT, Gunning RF, Hancock RD, Jeffrey M, Dawson M and Bradley M (1987) A novel progressive spongiform encephalopathy in cattle. Vet Rec 121:419–420

Wells GAH and Simmons MM (1996) The essential lesion profile of bovine spongiform encephalopathy (BSE) in cattle is unaffected by breed or route of infection. Neuropathol Appl Neurobiol 22:453

Wells GAH and Wilesmith JW (1995) The neuropathology and epidemiology of bovine spongiform encephalopathy. Brain Path 5:91–103

Wells G, Scott A, Wilesmith J, Simmons M and Matthews D (1994) Correlation between the results of a histopathological examination and the detection of abnormal brain fibrils in the diagnosis of bovine spongiform encephalopathy. Res Vet Sci 56:346–351

Wilesmith JW, Ryan JM, Atkinson MJ (1991) Bovine Spongiform Encephalopathy: Epidemiological Studies on the Origin. Vet Rec 128:199–203

Willoughby K, Kelly DF, Lyon DG and Wells GAH (1992) Spongiform encephalopathy in a captive puma (Felis concolor) Vet Rec 131:431–434

Wyatt JM, Pearson GR, Smerdon T, Gruffydd-Jones TJ, Wells GAH (1990) Spongiform Encephalopathy in a Cat. Vet Rec 20:513

Zlotnik I (1962) The pathology of scrapie:a comparative study of lesions in the brain of sheep and goats. Acta Neuropathol: Suppl 1:61–70

Zlotnik I, Rennie JC (1958) A comparative study of the incidence of vacuolated neurones in the medulla from apparently healthy sheep of various breeds. J Comp Path Therap 68:411–415

Public Health and the BSE Epidemic

M. N. Ricketts

Health Canada, Tunney's Pasture 0601E2, Building 6, Ottawa, Ontario, K1A 0L2, Canada

E-mail: Maura_Ricketts@hc-sc.gc.ca

Abstract Bovine Spongiform Encephalopathy was discovered in 1986 in the United Kingdom and relatively rapidly spread into its trading partners in Europe via contaminated cattle feed supplements. The practice of using the discarded bovine carcass as cattle feed supplements led to the recycling of the prion agent and the consequent generation of new point source epidemics in the recipient countries. The advent of rapid diagnostic tests and more widespread testing has led to the identification of BSE in countries not previously reporting cases and the recognition of larger numbers of infections in countries previously only reporting clinical cases. The recognition of the wider spread of BSE and the 1996 recognition of vCJD as a human disease caused by consumption of

BSE agent led to international concerns regarding the threat to human health and the demand for stricter controls on human food derived from cattle. Major shifts in food safety policy have occurred as a direct result. The recommendation that risk assessments for BSE infectivity and human exposure pathways be conducted rather than reliance upon rates and simple enumeration of BSE cases is one of the most prominent changes in the basis of policy regarding human health. The movement of BSE into human populations has a wider impact than seen in food safety—surgical procedures, blood, cells, tissues and organ donation programs are all affected. The World Health Organization has recommended that 'the eradication of BSE must remain the principle public health objective of national and international animal health control authorities'.

The opinions expressed in this chapter are those of the author and do not necessarily reflect the views of Health Canada. This review was written while the author was employed at the WHO.

1 The Global Spread of BSE

1.1 Introduction

The role of public health authorities in the development of policy against bovine spongiform encephalopathy (BSE) has not always been clear. The authority to prevent the spread of BSE within cattle populations and to control the importation of the agent through live animals, animal feed and food lies almost entirely within the function of veterinary health authorities. However, as it became clear that BSE is a zoonotic disease, routine involvement of public health authorities became inevitable. Interaction with public health authorities is necessary in order to develop appropriate national policy to protect human populations. The inclusion of public health authorities may help to address a particular concern of the public – that veterinary health policies have been perceived to be balanced in favor of the agrifood industry, lacking concern for human health. Unfortunately, whether this review is valid or not, the historical failure of national and regional authorities to develop appropriate policy for BSE led to extensive restructuring of government agencies, the creation of Food Safety Authorities in many countries and toppled Ministers.

Unlike many other important public health problems, testing and surveillance for BSE involves the examination of brain tissue, normally only available at death. As a result, screening for BSE is restricted to surveillance for clinical cases (not pre-clinical cases) and testing brain tissue in the slaughterhouse. Diagnostic tests for humans using tonsillar tissue, and advanced diagnostics such as magnetic resonance imaging (MRI) are available but are only useful in people who are already symptomatic. There is no test, as of September 2003, capable of detecting the agent in asymptomatic humans, in food, or in living, exposed animals. Nor is there a treatment known to prevent infection after exposure (prophylaxis), to prevent infection from becoming disease, nor to prevent disease from resulting inevitably in death.

The rapid and accelerating accumulation of new information has, in many cases, challenged the ability of national, regional and global authorities to develop appropriate responses. Additionally, public policy must be developed in the face of considerable ignorance as to the real level of risk in the nation. Policy developed from incomplete evidence, limited budgets and resources, cultural demands for safety, trade requirements and competing health care problems is challenging, but must be undertaken. As described below, there is good reason for all nations to examine their risk closely and to undertake at least some corrective action.

In this chapter I review BSE from the perspective of a public health physician. BSE is a complex disease, yet it also uniquely provides an opportunity for public health and veterinary authorities to develop relationships that will better serve the national interests, particularly in the overall area of food safety.

1.2 The Global Spread of BSE

The first cases of BSE to appear outside of the UK were reported from Ireland, followed closely by other trading partners in Europe. Importation of contaminated animal feed or infected animals led to the introduction of BSE and wherever rendering was practiced, infectivity was recycled into nationally manufactured meat and bone meal (MBM) supplements. Feeding the national herd with these supplements fueled the propagation of the national epidemic. Although the number of cases within Europe has remained relatively small (4323 reports as of 29 September 2003), control of the disease has eluded national authorities.

Table 1 Annual incidence rate (number of indigenous cases per million bovines aged over 24 months) of BSE in countries that have reported cases, excluding the United Kingdom

	Year													
Country	1989	1990	1991	1992	1993	1994	1995	1996	1997	1998	1999	2000	2001	2002
Austria	0	0	0	0	0	0	0	0	0	0	0	0	0.96	0
Belgium	0	0	0	0	0	0	0	0	0.61	3.69	1.84	5.53	28.22	25.75
Czech Republic	0	0	0	0	0	0	0	0	0	0	0	0	2.85	2.5
Denmark	0	0	0	0	0	0	0	0	0	0	0	1.14	6.77	3.35
Finland	0	0	0	0	0	0	0	0	0	0	0	0	2.39	0
France	0	0	0.45	0	0.09	0.27	0.27	1.09	0.54	1.64	2.82	14.73[a]	19.70	20.96
Germany	0	0	0	0	0	0	0	0	0	0	0	1.07	19.97	17.02
Greece	0	0	0	0	0	0	0	0	0	0	0	0	3.3	0
Ireland	4.41	4.12	5.00	5.14	4.57	5.43	4.57	20.28	21.39	20.79	22.83	38.17[a]	61.80[b]	88.39
Israel														6.25
Italy	0	0	0	0	0	0	0	0	0	0	0	0	14.1	10.6
Japan	0	0	0	0	0	0	0	0	0	0	0	0	1.44	.97
Luxembourg	0	0	0	0	0	0	0	0	10	0	0	0	0	14.54
Netherlands	0	0	0	0	0	0	0	0	1	1.01	1.03	1.07	10.25	13.19
Poland														1.28
Portugal	0	0	0	0	0	15.06	18.82	38.90	37.64	159.35	199.50	186.95	137.88	107.8
Slovakia	0	0	0	0	0	0	0	0	0	0	0	0	18.34	18.73
Slovenia	0	0	0	0	0	0	0	0	0	0	0	0	4.34	4.44
Spain	0	0	0	0	0	0	0	0	0	0	0	0.59	24.23	37.95
Switzerland	0	1	9.2	15.5	30.3	67.6	73.6	48.5	45.4	16	58.7	40.6	49.1	27.93

[a] France 2000: annual incidence rate in animals euthanized or found dead=5.45; annual incidence rate in BSE clinical cases=9.27; Ireland 2000: annual incidence rate in cases detected by the active surveillance program=17.93; annual incidence rate in BSE clinical cases=35.35.

[b] Ireland 2001: annual incidence rate in cases detected by the active surveillance program=29.90; annual incidence rate in BSE clinical cases=30.90. Reproduced by permission of the OIE (World organization for animal health), Paris, France, 2002. www.oie.int; last updated 3 June 2002.

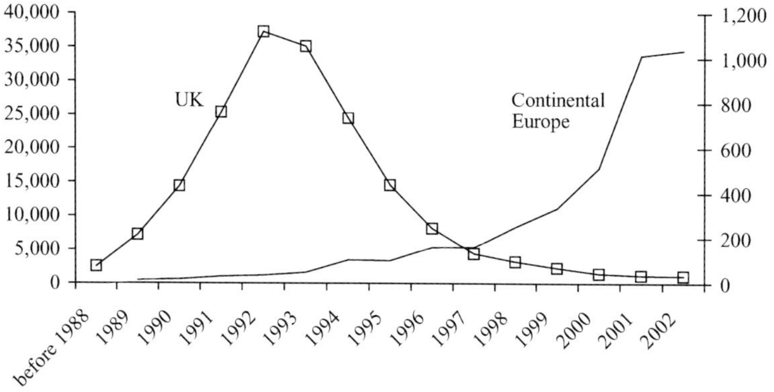

Fig. 1 Epidemic curve of BSE in Europe. UK, n=180,845; Continental Europe, n=3286. [Source, OIE (World Organization for animal health), Paris, France, 20 August 2002]

Until recently the number of cases and the incidence of disease had been increasing (Fig. 1 and Table 1). Thirteen countries reported their first cases of BSE after the year 2000 (Table 2).

It is clear that some countries were aware of their risk from imported MBM and live bovines, but did not take action until the first clinical cases of BSE were identified. Early, passive, surveillance systems depended

Table 2 BSE reports from countries reporting first cases after 2000

Country	Number of cases reported 2000–2003
Austria	1
Canada	1
Czech Republic	5
Finland	1
Germany	273
Greece	1
Israel	1
Italy	87
Japan	7
Poland	8
Slovakia	12
Slovenia	3
Spain	325

Source: Office International des Epizooties (OIE) (World organization for animal health), Paris, France, 2002, www.oie.int. Last updated 29 September 2003.

upon the identification and reporting of clinical cases. However, active surveillance uses rapid tests on the brains of cattle, and directed surveillance toward emergency slaughter and downer cattle. A key policy change in the European Community (EC) in 2001 legislated the use of active surveillance. This strategy identified BSE in countries where cases had not been previously reported, and doubled the number of reports from others.

1.3 Tracing the Movements of BSE Globally

After April 1985, when the first BSE case was identified in the UK, the UK was still legally exporting MBM as well as live bovines, offals (potentially containing higher risk tissues) and various food products. It is important to note that these exports were legal, and that the UK was undertaking a series of increasingly meticulous activities that progressively removed larger amounts of infectivity from the feed and food chain in the UK. Exported MBM feed supplements were labeled 'not for use in ruminant animals'. In addition, it was not clear in the early years of the epidemic that BSE was a human health threat, although the issue was recognized and commented upon, with concern, within the Southwood Committee and by some scientists. It was 28 July 1989 before the EC banned the import from the UK of 'cattle born before July 1988 and offspring affected or suspected animals' [1]. On 13 November 1989, England and Wales banned the use of certain specified bovine offals (SBO) for human consumption [2]. The EC, on 9 April 1990 acted to formalize a decision to ban imports of SBO and other tissues from the UK [3]. In 1990, the UK extended the ban on the use of SBO in any animal feed in the UK [4]. Exports of such feed was also effectively banned to other member states.

Nonetheless, the ability of the original SBO bans to remove all infectivity can be questioned. The list of SBO included those tissues recognized at the time as being heavily contaminated with the prion agent, but this list was increased over time to include the whole head, spinal cord and then spinal column, entire gut, lymph nodes and some organs. In addition, inspection was not as rigorous in the earlier years of the epidemic as it became later, particularly after it was realized that even very small amounts of brain tissue could infect cattle. While the removal of SBOs from cattle carcasses was successively improved over time, particularly in the UK, it was certainly incomplete. A small number of cases of

BSE continue to occur in the UK, indicating that infectivity had been present in animal feed until recent years.

In any case, third country exports of feed from the UK were not banned under legislation until 10 July 1991. In 1996 the EC enacted an embargo against the UK preventing exports of all bovine products, and has since taken substantial action against BSE throughout the EC. However, by this time BSE had become established in a number of countries, and these countries were also exporting MBM, live bovines, offals and bovine based foods beyond their borders. Other countries with BSE may not yet have adopted the rigorous methods used in the UK.

2 Risk Assessment

The first step in the determination of appropriate public policy is conduct of a risk assessment. The risk assessment must address the question of whether there is a risk of BSE, not simply whether BSE is present or not. The Joint World Health Organization (WHO)/Food and Agriculture Organization (FAO)/Office international des epizooties (OIE) Consultation on BSE: public health, animal health and trade [5], specifically commented that a country cannot determine if it must take action against BSE based simply upon the rate or number of cases of BSE in a country. As a result, the OIE Terrestrial Animal Health Code (the Code) [6] is un-

Table 3 Risk based recommendations for export

Country considered as BSE free:
May export without restriction
Country with minimal BSE risk:
May not export brains, eyes, spinal cord, skull, vertebral column and derived protein products from cattle aged over 30 months
Country with moderate BSE risk:
May not export brain, eyes, spinal cord, distal ileum, skull, vertebral column and derived protein products from cattle aged over 6 months
Country with high BSE risk:
May not export brain, eyes, spinal cord, tonsils, thymus, spleen, intestines, dorsal root ganglia, trigeminal ganglia, skull, vertebral column and derived protein products from cattle aged over 6 months

Source: OIE Terrestrial Animal Health Code.

dergoing review to become more clearly risk based. The Code informs the WTO in the event of a trade dispute (Table 3). In addition, the consultation recommended that risk management must be commensurate with the level of risk. Based upon the risk assessment, some countries will need to do more than others. The consultation further recommended that risk assessment methodologies should consider the model developed by the EC.

2.1 Geographic Based BSE Risk Assessment

The EC has prepared a BSE risk based assessment methodology through ad hoc BSE committees reporting to the Scientific Steering Committee of the European Commission. The Geographical Risk of Bovine Spongiform Encephalopathy (GBR) models the risk within a country by first determining whether, or how much, BSE has been introduced into the country [7]. Data regarding imports from the UK and other European countries (the EC has recently decided to amend its GBR to include an assessment from all countries currently graded GBR III) is used. The model then requires a description of the use made of any bovines or bovine based products in order to determine if there was any risk of recycling within the cattle population (i.e., are cattle consuming MBM; is

Fig. 2 Map of the world showing GBR classification

there a rendering industry and recycling). Finally, any mitigating factors are considered.

This method has demonstrated its predictive capacity in that some countries were classified as GBR III before the detection of their first cases (Germany, Italy, Spain, Czech Republic and Slovak Republic). Denmark, Greece and Japan had their first case before formal completion of their assessment; indicators were that they would be graded GBR III.

The model provides, in addition to a categorization into one of four risk levels (Fig. 2, Tables 4, 5), an estimate of the stability, i.e., future course of the epidemic. Any country wishing to trade products of bovine origin into the EC must submit to this evaluation.

Table 4 Definition of GBR and its levels

GBR level	Presence of one or more cattle clinically or pre-clinically infected with the BSE agent in a geographical region/country
I	Highly unlikely
II	Unlikely but not excluded
III	Likely but not confirmed, or confirmed at a lower level
IV	Confirmed, at a higher level

Source: GBR (EC/SSC)

Table 5 All countries with A GBR classification

GBR I
Argentina, Australia, Botswana, Brazil, Chile, Costa Rica, El Salvador, Iceland, Namibia, New Zealand, Nicaragua, Norway, Panama, Paraguay, Singapore, Swaziland, Uruguay, Vanuatu
GBR II
Canada, Colombia, India, Kenya, Mauritius, Nigeria, Pakistan, Sweden, USA
GBR III
Albania, Austria, Belgium, Bulgaria, Croatia, Cyprus, Czech Republic, Denmark, Estonia, Finland, France, Germany, Greece, Hungary, Ireland (Republic), Italy, Japan, Latvia, Lithuania, Luxembourg, Netherlands, Poland, Romania, San Marino (Republic of), Slovak Republic, Slovenia, Spain, Switzerland, Turkey
GBR IV
Portugal (mainland), United Kingdom

Source: EC Table based upon the Update of the GBR-opinion of the SSC of 11 January 2001, and subsequent assessments of Vanuatu, Turkey, Republic of San Marino, Latvia, Iceland, Croatia, Bulgaria, Finland, Austria.

2.2 Preventing and Controlling BSE

The catalogue of interventions is well understood. Documents such as the Proceedings of the Joint WHO/FAO/OIE Technical Consultation on BSE: public health, animal health and trade, the European Community legislation and advice of the Scientific Steering Committee [8] and documents prepared by the UK authorities [9, 10, 11, 12] provide substantial advice. The OIE Code (OIE) provides guidance regarding trade, and in doing so provides the requirements for a BSE risk assessment.

Briefly, all countries must know the epidemiology for all animal transmissible spongiform encephalopathies (TSEs) in the country, with surveillance and culling programs. Surveillance for BSE must use the newer rapid tests (active surveillance) and not be dependent upon clinical diagnosis. In addition, the appropriate population of bovines must be sampled, such as older cattle and downer cows. Focusing on neurological disease has not proven to be sufficiently reliable. All countries must review the importation of potential BSE sources including live animals and bovine-based food and feeds. Slaughter techniques must be reviewed. In countries where there is a risk of BSE, measures must be taken to avoid contaminating the carcass with high risk tissues during killing (e.g., slaughter technique such as captive bolt) or butchering (e.g., if the carcass is split through the spinal column there may be contamination of the carcass). The removal of high risk tissues (also referred to as specified risk materials—SRM) contributes immediately to improved human and animal safety. In countries with any risk of BSE, those tissues known to have infectivity must be removed. In countries with a high risk of BSE, even tissues that are suspected of carrying BSE infectivity must be removed. The removal of the entire head and spinal column may be necessary. Feed bans must be implemented, as bovines must not be exposed to any tissue that contains BSE infectivity. The WHO called for a worldwide ban on ruminant tissues in ruminant feed in 1996 [13]. However, implementation must be carefully evaluated before concluding that such a ban has actually been enforced, as cross-contamination by exquisitely small amounts of tissue (a piece of brain tissue smaller than a peppercorn) can infect a bovine. In the UK it became necessary to impose a ban on the feeding of rendered protein products to all farm animals because bovines continued to be exposed to minute quantities of infectivity through other animal feed in the barnyard. Some countries at low risk have identified selected proteins of bovine origin (such as dried blood or table scraps) that are still used. In countries continuing to feed their bovines with rendered meat and bone meal supplements, or where avoidance of cross-contamination between animal feed cannot be guaranteed, treatment processes must be able to reduce infectivity, particularly in the production of MBM.

It is possible for a country that has not conducted a risk assessment and lacking adequate surveillance to be manufacturing food with a higher intrinsic risk than countries with excellent surveillance systems, a known rate of BSE and control measures whose implementation is being rigorously assessed.

3
BSE and variant Creutzfeldt-Jakob Disease: Risk for Human Populations

3.1
Risks and Hazards

Public demands for food safety are influenced by a legacy of mistrust that began in many countries with the BSE crisis. The transmission of hepatitis C virus (HCV) and human immunodeficiency virus (HIV) through transfusion and scandals regarding hospital and health care provider failures contribute to a general skepticism regarding the competence of public authorities. Unfortunately, particularly with BSE, announcements and scientific breakthroughs can arrive with alarming rapidity, requiring a flexible and iterative policy development process. The appearance of ignorance and flip-flop policy development can be mitigated through open and regular interaction with the public (possibly through consumer groups) throughout the process of policy development. Some risks are well understood, others less so (Table 6). When moving from 'known risks' to 'hazards' and 'hazards influenced by risk perception', there is an accompanying shift in evidence. In addition, it must be the directly stated goal of public policy that the health of humans is of greatest importance. This is particularly important in areas where the current policies are pre-emptive in nature, i.e., developed for theoretical risks, such as is the case for current blood policy regarding the TSEs. In any case, the general public may demand very stringent

Table 6 Risks and hazards

Known risks

The measures required to avoid epidemic BSE, described in Sect. 2.2, are well understood

Level of intervention is based on:

- Risk assessment (refer to Joint Technical Consultation of WHO/FAO/OIE, GBR and OIE International Animal Health Code)
- Trade requirements (refer to EC and OIE trade requirements)

Deduced hazards

It is assumed that humans became exposed to BSE through food from infected or contaminated bovine tissues

There are other products of bovine origin that are known to contain high risk tissues, or where the route of exposure is through injection or transplantation e.g.:

- Biopharmaceuticals and transplantable tissues made from or involving bovine tissues during their production
- Herbal medications using bovine (and ovine) tissues
- Medicines, biological products and organs from persons with vCJD
- Surgical procedures in countries where vCJD has been identified
- Blood safety
- Cosmetics

Sheep and goats are susceptible to BSE and although BSE has not been found in national herds, attention must be paid to avoiding exposure to tissues that might harbor BSE infectivity, whenever exposure may have occurred

Hazards influenced by risk perception

Environmental contamination, i.e.:

- By-products of incineration of cattle carcasses
- Storage of MBM
- Contaminated water run-off
- Nonfood products such as fertilizer

Products made from bovine tissues not currently known to carry BSE infectivity, such as muscle

Bovine products that are heavily processed or that contain very small amounts of high risk bovine tissues may have no risk yet be regarded with concern i.e., gelatin or beef bouillon

safety interventions even where there is no certainty that safety will be improved.

Among scientists there is a general tendency to believe that decision making should be based exclusively upon science. In reality, the process by which a member of the general public makes decisions about risk is markedly different from that of scientists. Risk perception describes the

Table 7 Risk perception and vCJD

New disease
Agent not thoroughly characterized
Research facilities are high security, supporting the impression that BSE is contagious and dangerous
The processes (such as rendering cattle carcasses to make cattle feed) are poorly understood and seem 'unnatural'
Discovery is ongoing, so recommendations are always changing, leaving the impression that no one knows what they are doing
Long, hidden incubation period; there is no test to determine if there has been exposure in asymptomatic people
vCJD always results in death, and the disease process is highly dreaded—mute, blind, incontinent, immobile/paralyzed/bedridden, uncertainty about level of consciousness and pain
The public feels that the individual is forced to take a risk so that an industry can protect its profit margin

factors influencing a member of the general public when they are deciding whether or not to take a risk (Table 7). As an example of the influence of risk perception on risk management, a scientific decision may be made that mechanically recovered meat can continue to be collected, possibly because the risk of BSE is considered to be very low. When incorporating knowledge about risk perception into policy making, the policy of avoiding human exposure to neurological tissues may require that mechanical crushing of bones to recover meat paste be prohibited. The final policy decision may recommend that meat recovery systems must mitigate or entirely exclude neurological tissue.

3.2 Exposure to BSE

Evaluating the risk for any individual who has been exposed may be impossible. The task is complicated by the inherent difficulties of epidemiologic investigation using food histories and the absence of historical samples to test for infectivity. Most importantly, there is insufficient knowledge regarding the dose of BSE that will infect a human. It is not yet understood why one person in a family fell victim to the disease while the others remained unaffected. Because of this uncertainty, it is appropriate to develop a policy in which humans are not exposed to any tissues known or suspected to contain BSE.

3.2.1
Human Exposure Pathways

Assessment of the human exposure pathways will describe all the ways by which humans can be exposed to BSE, whether through food, pharmaceuticals or other products. Relevant human exposure pathways are likely to differ between countries, hence each country must internally review the fate of high risk tissues whenever there is a risk of human exposure to BSE from the national herd, and must evaluate external sources of exposure.

The number of products containing bovine tissues is myriad. More than 50% of the carcass of a bovine cannot be used for human food directly. Through rendering, the bovine carcass is converted from a hazardous waste into a wide range of useable products. The products include protein meals (i.e., MBM based animal feeds), proteins (peptides and amino acids), tallow (used in food and nonfood uses such as in the oleochemical industry for soap and fatty acid derivatives), rendered fats (used as milk replacers for young ruminants), gelatin, rennet (used in the production of human and animal foods, i.e., cheese and food supplements, in medicines and lactose production), and in the production of phosphates from bone and collagen. Bovine materials are used in the manufacture of medicines (biopharmaceuticals), blood products, biologics (such as dura mater), herbal medications and cosmetics. The measures to ensure safety (such as control over the health of the herds or which tissues are used) vary widely among these products.

3.2.2
Dose

As of September 2003, approximately 136 cases of variant Creutzfeldt–Jakob disease (vCJD) had been reported worldwide. Given the wide geographic dispersion of potentially contaminated food and food products from countries with BSE over at least a decade, it is interesting that there have been so few cases. Some researchers have suggested that the dose required to infect humans is quite high and so only a few persons received a dose above the threshold for human infection. However, if the incubation period is very long, more cases will be forthcoming.

The BSE agent is not distributed evenly throughout the body of the bovine animal. By focusing attention on avoiding exposure to high risk

Table 8 Table of BSE distribution in bovine animals (EC/SSC)

Tissue	Cattle infectivity dose $(ID)_{50}$ per BSE case	Percentage of total infective load per bovine
Brain	5,000	64.1%
Spinal cord	2,000	25.6%
Trigeminal ganglia	200	2.6%
Dorsal root ganglia	300	3.8%
Ileum	260	3.3%
Spleen[a]	26	0.3%
Eyes	3	0.04%

[a] The EC SSC hypothesized that the spleen could contain some infectivity based on other TSEs. However, no such infectivity has been found after intracerebral inoculation of cattle.

tissues, considerable safety can be conferred. Table 8 shows the distribution of BSE infectivity in bovine animals.

Many measures taken to reduce the risk that a bovine can be exposed to BSE will clearly have direct impact on whether humans are exposed to tissues or products containing BSE. The Joint WHO/FAO/OIE Technical consultation on BSE: public health, animal health and trade, recommended that the whenever a country identifies a risk for BSE, it should take steps to identify the fate of SRM. It is notable that exposure to these tissues can occur even in the absence of actual exposure to them if slaughter methods permit contamination of tissues otherwise regarded as containing no infectivity.

3.2.3
Factors Affecting Infectivity of Tissues

The manufacturing processes for each product must be investigated, as some processes may be able to wholly or significantly remove infectivity. It should be noted that processes such as cooking at temperatures used in the routine manufacture of food is insufficient to remove infectivity from BSE contaminated products. Exposure to hot alkaline (bleach) or batch processing at temperatures no less than 133°C at 3 bar pressure for 20 min (recommended by EC for rendered animal tissues in the EU) are among the few methods known to remove BSE infectivity. Dilution must be considered, but it is possible that humans accumulate their risk, and that dilution does not protect again infection if exposure is ongoing.

3.2.4
Route of Exposure

As stated earlier, the most likely route of exposure for humans who have developed vCJD is through BSE contaminated food. Traditionally, this is a relatively lower risk route of exposure, following intracerebral, implantation, inoculation and intravenous exposure, all considered to be higher risk. Oral exposure is higher risk than either topical or inhaled routes of exposure. Neither topical (except maceration of dental tissues in experimental animals) nor inhaled routes of exposure have been associated with transmission of any TSE.

Nevertheless, when evaluating risks to humans, it is necessary to consider in what ways humans could be exposed through all routes. It can be expected that the dose that can infect a human through intracerebral, implantation/transplantation, inoculation or intravenous routes, particularly if there is direct exposure of the brain, would be smaller by orders of magnitude than an oral exposure. For this reason, there must be a lower tolerance for invasive exposures. Some authorities may have lower tolerance for nonessential exposures, i.e., cosmetics. Finally, there is generally lower tolerance for exposures where medical trust is involved, i.e., surgical, medicines, transplantation, and biologics including vaccines.

3.2.5
What Is Safe to Eat?

In any case, the OIE Code recommends that milk and milk products, semen, protein-free tallow, dicalcium phosphate, hides and skins, as well as gelatin and collagen prepared from hides and skins should be traded without restrictions. The Code provides specific requirements regarding trade in live animals and other bovine products, including fresh meat. However, it is considered by many, including the WHO and the Scientific Steering Committee (SSC) of the EC, that skeletal muscle meat free of visible nervous tissue can be eaten with as much confidence as milk products. Skeletal muscle meat is not known to carry infectivity; none has yet been detected despite numerous (ongoing) investigations. The gelatin manufacturers have sponsored a series of investigations and feel that their product may be essentially safe [15]. Many other products have been reviewed by the SSC of the EC and the results of these reviews are available on their website [14]. Other national authorities may also make their risk assessments available upon request.

In recognition of the complexity of food production, the Joint WHO/FAO/OIE Technical consultation on BSE: public health, animal health and trade, called for a standardized international approach to identify whether SRM was included in foods distributed through international trade.

3.3 Exposure to vCJD

Finally, it is necessary to evaluate the risks from exposure to a person with vCJD. The human TSEs are not, in fact, easily transmitted. Transmission is known to have occurred only through the exposure to tissues either extracted from the brain or which were in immediate contact with the brain (cornea, dura mater and human cadaver-sourced growth hormone and gonadotropin), or through neurosurgical exposures. It has never been transmitted between humans through care of a person with a TSE, through cutaneous exposure, needle-stick injuries (even of brain tissue in experimental settings) nor are respiratory routes of exposure considered to be a risk. Transfusion exposures have not led to cases of sporadic CJD.

Among humans with vCJD there is limited information about the distribution of the agent. However, all tissues of the central nervous system, including the eye, are considered to be infected. In addition, for the purposes of infection control other tissues may be regarded as having infectivity based upon animal models (Table 9, Proposed Infectivity Category for Infection Control).

In vCJD, tonsils (EEC) and lymph nodes [16, 17] have been identified to contain prion protein. Population studies in the UK have not found evidence of prion protein among persons who are not ill with the disease [18]. In countries with endemic cases of vCJD, these findings necessitated a review of infection control (particularly the management of surgical instruments) [19] and human tissue sourcing guidelines, as well as therapeutic products derived from human tissues [20]. Variant CJD has had a profound impact on the blood industry.

Table 9 Proposed infectivity category for infection control (WHO 1999)

Infectivity category[a]	Tissues, secretions, and excretions	
High infectivity	Brain	
	Spinal cord	
	Eye	
Low infectivity	Cerebrospinal fluid	
	Kidney	
	Liver	
	Lung	
	Lymph nodes/spleen	
	Placenta	
No detectable infectivity	Adipose tissue	Feces
	Adrenal gland	Milk
	Gingival tissue	Nasal mucous
	Heart muscle	Saliva
	Intestine	Semen
	Peripheral nerve	Serous exudate
	Prostate	Sweat
	Skeletal muscle	Tears
	Testis	Urine
	Thyroid gland	
	Blood[b]	

[a] Assignment of different organs and tissues to categories of high and low infectivity is based chiefly upon the frequency with which infectivity has been detectable, rather than upon quantitative assays of the level of infectivity, for which data are incomplete. Experimental data include primates inoculated with tissues from human cases of CJD, but have been supplemented in some categories by data obtained from naturally occurring animal TSEs. Actual infectivity titers in the various human tissues other than the brain are extremely limited, but data from experimentally infected animals generally corroborate the grouping shown in the table.

[b] Experimental results investigating the infectivity of blood have been conflicting, however even when infectivity has been detectable, it is present in very low amounts and there are no known transfusion transmissions of CJD. This document will classify these tissues as having no detectable infectivity ('no detectable infectivity tissues') and, for the purposes of infection control, they will be regarded as noninfectious

4 Conclusions

The global BSE risk is not easy to quantify, but through the mechanism of trade, wide geographic areas of the world have had ample opportunity to import BSE infectivity to within their borders. It is impossible to determine the origin of all food and feed products, because, in general, international packaging lists only the country of manufacture of the product. There is no international obligation to list the country of origin of all of the components. If one also considers movements due to illegal repackaging and trade, the distribution of the agent may be broader than is currently understood.

The need for all countries to conduct risk assessments and to undertake appropriate action is urgent. The Joint WHO/FAO/OIE Technical Consultation on BSE: public health, animal health and trade, identified countries in Eastern Europe (Czech Republic, Poland, Slovakia, Slovenia have now identified BSE cases), the middle East and North Africa (Israel has since reported its first case), and in South East Asia (Japan has reported cases of BSE) as being at risk from BSE due to their importation of substantial quantities of MBM from the UK. Analysis of the overall patterns of exportation from GBR III countries has not been published.

At a global level it is clear that despite the knowledge of the measures necessary to control BSE, they are not being implemented. In 2002, the OIE International Committee invited all OIE Member Countries to carry out an assessment of the risk of BSE being present in its domestic cattle herd. There may be a reluctance to disrupt existing agricultural and animal husbandry practices in countries where BSE cases have not yet appeared. Yet the lesson from the last decade is that wherever there is a risk, some measures must be taken. Where there is a high risk that BSE contaminated bovines or bovine products have been imported into a country, there is a strong obligation to undertake corrective measures.

The measures undertaken within the EC required resources, organization and financial elasticity that are simply not present in some parts of the world. The commercial impact of eradication of BSE can be enormous, affecting the economic stability of a country. At the Joint WHO/FAO/OIE Technical Consultation on BSE: public health, animal health and trade, estimates were that between March 1996 and 2001, the cost of BSE in the UK was approximately £4 billion. Beef consumption in Europe has dropped at times by 40%.

Taking action against BSE may be essential to protect trade in bovine and ovine products, as the appearance of a single BSE case can result in immediate embargoes on exportation of bovine products from a country. For these reasons it is essential that national and international authorities move quickly; prevention is better than cure. If a country is heavily dependent upon cattle and cattle products for trade related income, the implementation of the necessary measures to eradicate the disease may have to be taken at the expense of other programs. However, it should be recognized that the establishment of BSE in the national herds of some countries may be, practically speaking, too expensive for the national authorities to eradicate.

While it can be anticipated that considerable expertise, or at least the capacity to develop the expertise, to prevent the spread of BSE will lie within veterinary health authorities, there is an important need for public health officials to conduct a risk analysis that is oriented toward public health issues. Public health practitioners will need to use their knowledge to influence national public health and veterinary policies. In many countries public health officials are involved in food safety only when there is an outbreak of a food-borne disease. Hence to influence veterinary practices, they must become knowledgeable about food safety and the entire food chain. Public health officials may be warranted in recommending policies that extend beyond those required for animal health. In addition, public health practitioners are familiar with the interaction between scientific decision making regarding risk and the needs of the population for high levels of food security and health safety.

The WHO Consultation on Public Health and Animal TSEs: Epidemiology, risk and research requirements [21] concluded that 'the eradication of BSE must remain the principle public health objective of national and international animal health control authorities'.

References

1. EEC Decision 89/469/EEC, 1989
2. Bovine Offal (Prohibition) Regulations 1989 (SI 1989 No 2061), UK
3. EEC Decision 90/200/EEC, April 1990
4. Bovine Spongiform Encephalopathy (No.2) Amendment Order 1990 (SI 1990 No 1930), UK
5. Proceedings of the Joint WHO/FAO/OIE Consultation on BSE: public health, animal health and trade. Paris, France. June 2002
6. OIE Terrestrial Animal Health Code Chapter 2.3.13.1 12th Edition, May 2003 Bovine Spongiform Encephalopathy http://www.oie.int/eng/normes/en_mcode.htm, 2003
7. Opinion of the Scientific Steering Committee on the Geographical Risk of Bovine Spongiform Encephalopathy (GBR) SSC/11/01/2002/6.2.c.1 adopted 11 January 2002

8. Opinions and Reports from the Scientific Steering Committee http://europa.eu.int/comm/food/fs/bse/index_en.html, 2003
9. UK Food Safety Agency Review of BSE Controls http://www.bsereview.org.uk/data/final/final_contents.htm
10. Department of Environment, Food and Rural Affairs (DEFRA) Publications on implementation and regulation http://www.defra.gov.uk/animalh/bse/, 2003
11. Department of Environment, Food and Rural Affairs (DEFRA) Reports from the Spongiform Encephalopathy Advisory Committee http://www.defra.gov.uk/animalh/bse/public-health/public-health-index.html
12. Report of the UK BSE Inquiry October 2000, written by the Members of the Committee of the Inquiry, Lord Phillips of Worth Matravers, Mrs June Bridgeman, Professor Malcolm Ferguson-Smith. http://www.bseinquiry.gov.uk/, 2003
13. Report of a WHO Consultation on Public Health Issues related to Human and Animal Transmissible Spongiform Encephalopathies. Document WH/EMC/DIS/96.147. Geneva, Switzerland 2–3 April 1996
14. Opinion of the Scientific Steering Committee: Listing of Specified Risk Materials, a scheme for assessing relative risks to man: adopted on 9 September 1997, re-edited version adopted by the Scientific Steering Committee during its third Plenary Session of 22–23 January 1998
15. Report of the Gelatine Manufacturers Association, 2003, www.gelatine.org
16. Hill AF, Butterworth RJ, Jointer S, et al: Investigation of variant Creutzfeldt-Jakob and other human prion diseases with tonsil biopsy samples. Lancet 353:183–199, 1999
17. Hilton DA, Fathers E, Edwards P, Ironside JW, Zajicek J. Prion immunoreactivity in appendix before clinical onset of variant Creutzfeldt-Jakob disease. Lancet 352:703–704, 1998
18. Ironside JW, Hilton DA, Ghani A, Johnstone NJ, Conyers L, McCardle LM, Best D. Retrospective study of prion-protein accumulation in tonsil and appendix tissues. Lancet 355 (9216): 1693–1694, 2000
19. Report of the WHO Consultation on Infection Control Guidelines for Transmissible Spongiform Encephalopathies WHO/CDS/CSR/APH/2000.3 Geneva, Switzerland 24–26 March 1999
20. WHO Guidelines on Transmissible Spongiform Encephalopathies in Relation to Biological and Pharmaceutical Products. WHO/BCT/HTP Geneva, Switzerland, 3–5 February 2003
21. Report of the WHO Consultation on Public Health and Animal TSEs: Epidemiology, risk and research requirements Geneva 1–3 December 1999

Clinical Features of Variant Creutzfeldt–Jakob Disease

R. G. Will · H. J. T. Ward

National Creutzfeldt-Jakob Disease Surveillance Unit, Western General Hospital, Edinburgh, UK

E-mail: r.g.will@ed.ac.uk

Abstract The possibility that a new form of human prion disease, variant Creutzfeldt-Jakob disease (vCJD) had occurred in the UK was first raised by the identification of a small number of cases with unusual clinical characteristics. Atypical features included a young age at death, a predominantly psychiatric presentation, a relatively extended duration of illness and the absence of the 'typical' periodic electroencephalogram seen in sporadic CJD. Diagnostic criteria for vCJD have now been formulated and partially validated. Magnetic resonance imaging of the brain shows high signal in the posterior thalamus in the great majority of cases and all tested cases to date have been methionine homozygous at codon 129 of the prion protein gene (*PRNP*). There is a need to try and improve early diagnosis, particularly if effective treatments are developed.

1 Introduction

The UK National Creutzfeldt–Jakob Disease Surveillance Unit identified variant Creutzfeldt–Jakob disease (vCJD) as a novel form of human prion disease in 1996 (Will et al. 1996). The original hypothesis that vCJD was caused by transmission of bovine spongiform encephalopathy to the human population has been supported by subsequent epidemiological evidence and by laboratory research, including transmission experiments in rodent models. The possibility that a new form of CJD had developed in the UK was initially raised by the identification of a small number of cases of CJD with unusual clinical characteristics, which included a young age at death and a predominantly psychiatric presentation. Analysis of larger numbers of cases has led to the identification of a consistent clinical phenotype (Will et al. 2000), which is relatively distinct from other forms of human prion disease, including sporadic CJD (sCJD). This article reviews the clinical features of vCJD.

2 Age at Death in vCJD

The most striking characteristics of vCJD is the relatively young age at death in comparison to sCJD (Fig. 1). The mean age at death in vCJD is 29 years (range, 14–74 years) in comparison to a mean age at death of 66 years (range, 20–95 years) for sCJD in the UK since 1990. There is an

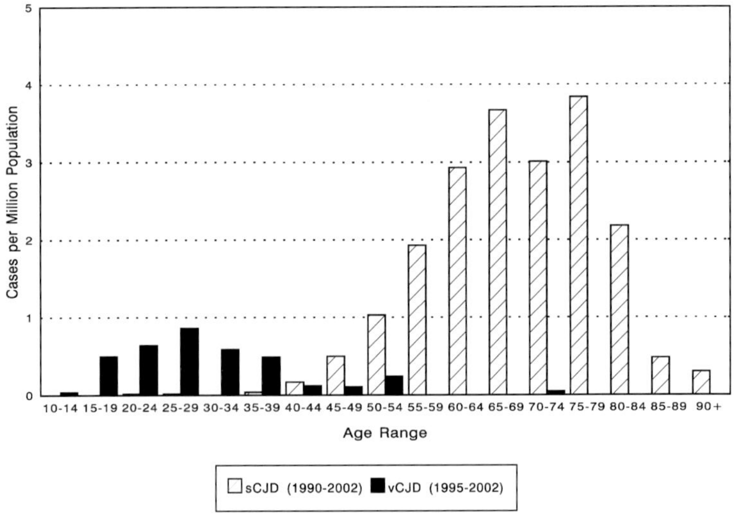

Fig. 1 Mortality rates by 5-year age groups in sCJD and vCJD

overlap in the age distribution in the two conditions and even in the 40–49-year age group sCJD has a higher incidence than vCJD. This is of importance in that age by itself cannot be used as a criterion to distinguish between the vCJD and sCJD. The suspicion of a diagnosis of vCJD may be raised by the development of suggestive features in a younger person, but a probable diagnosis depends on an assessment of clinical characteristics, with a definite diagnosis resting on neuropathological confirmation (Ironside et al. 2000).

Identifying the true age distribution in CJD of any type assumes a high degree of case ascertainment across the age range. In sCJD there has been concern that cases in the elderly population may be missed and in the UK between 1980 and 1995 an apparent doubling of the incidence of sCJD is largely attributable to an increase in the numbers of elderly cases, probably due to improved case identification in this age group. In vCJD a separate system for the identification of cases in the paediatric age group was established in the UK in 1996 and six cases of vCJD have been identified with an age at the onset of clinical symptoms of less than 16 years (Verity et al. 2000). Importantly the clinical and pathological characteristics of these cases are identical to those of older cases. The case of vCJD dying at the age of 74 years (Henry et al. 2002) is an outlier and there is uncertainty about whether cases of vCJD in the elderly may be missed by current surveillance methodologies. The UK surveillance system uses death certificates as a means of identifying cases of vCJD not directly notified and no 'missing' cases of vCJD in the elderly have been identified by this mechanism.

3
Clinical Features of vCJD

The initial symptoms in vCJD are usually psychiatric, most frequently depression, anxiety and withdrawal, although a minority of cases exhibit first-rank symptoms suggestive of a psychotic illness. After a median of 6 months clear-cut neurological features develop, including cognitive impairment, ataxia and involuntary movements. Chorea and dystonia occur as well as the myoclonic movements which are a common characteristic of sCJD. The clinical course is relentlessly progressive with the development of dementia and diffuse cortical deficits. Terminally patients are usually mute, bed-bound and helpless. Death occurs a median

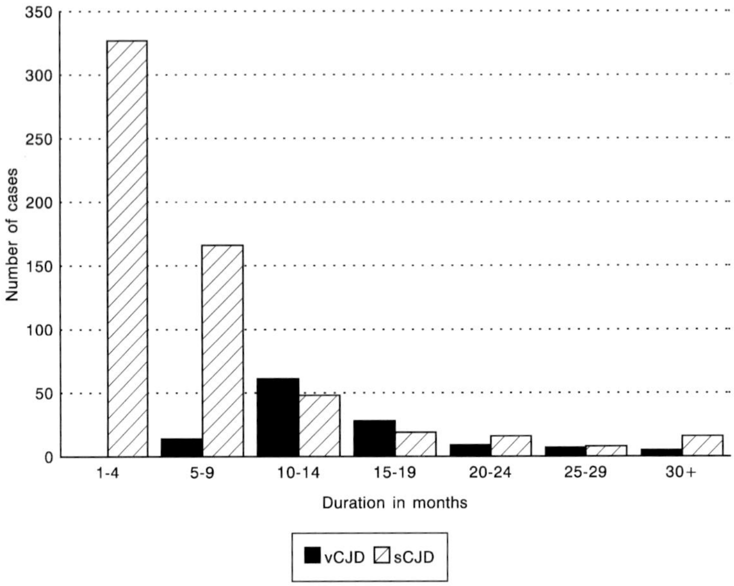

Fig. 2 Duration of illness in sporadic and variant CJD

of 14 months from the onset of symptoms (range, 6–39 months) and is often due to an intercurrent infection such as bronchopneumonia.

The total duration of illness from first symptom to death is an important parameter in the classification of neurological disease. Although there is an overlap in the duration of illness in vCJD and sCJD (Fig. 2), there are no cases of vCJD with an illness duration of less than 6 months. This is clinically important because the majority of cases of sCJD present subacutely with rapidly progressive dementia and deteriorate rapidly to a state of akinetic mutism (Brown et al. 1986). In about 5% of cases of sCJD the presentation is so acute that the diagnosis of stroke is suspected (McNaughton and Will 1997). In contrast vCJD is characterized by an initial phase, lasting months in which the features are predominantly psychiatric and even if there are early neurological or cognitive symptoms these progress only slowly in the initial stages (Zeidler et al. 1997b). The speed of evolution of neurological deficits is therefore useful in distinguishing between sCJD and vCJD. However, the phenotype in sCJD is variable and there are subgroups of sCJD with a clinical picture similar to vCJD, including a young age at onset and prolonged illness du-

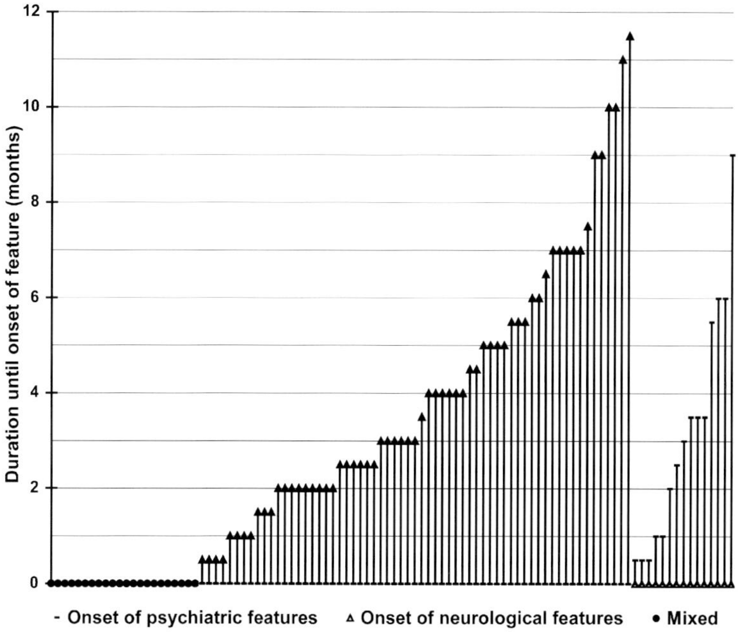

Fig. 3 Case-by-case plots of onset of psychiatric and neurological features (cases sorted by times to onset)

ration (for example cases with a valine/valine genotype and deposition in the brain of the type 1 isoform of prion protein) (Parchi et al. 1999). It may be impossible to classify cases of CJD solely on the basis of the clinical features.

An analysis of the neurological and psychiatric features of the first 100 cases of vCJD has been published recently (Spencer et al. 2002) with the aim of clarifying the early clinical characteristics of vCJD. One important conclusion of this study is that the clinical presentation of vCJD is heterogeneous: 63% of cases present with psychiatric symptoms alone, 15% with isolated neurological symptoms and 22% with mixed features (Fig. 3). In the group with initial neurological symptoms 11/15 cases developed associated psychiatric symptoms within 4 months of onset and all cases had mixed neurological and psychiatric symptoms by 9 months of onset.

In the cases with a purely psychiatric mode of onset the most frequent features were behavioural change, dysphoria, irritability, anergia, withdrawal and anxiety. However, a significant minority of cases (10/63) had poor memory as an early symptom. In the cases with a neurological onset there were a range of symptoms, but sensory features were predominant and included pain, paraesthesia, numbness, hypersensitivity, coldness and 'odd sensation'. Four cases exhibited gait disturbance as an initial symptom and two slurring of speech. In the mixed category there were a wide range of neurological and psychiatric symptoms, but again pain or sensory symptoms were prominent in a significant proportion of cases. Gait disturbance was a feature in six, poor memory in three and slurring of speech in one.

This analysis together with a previous studies of the subsequent evolution of psychiatric and neurological symptoms (Will et al. 1999, 2000; Zeidler et al. 1997a) suggests that, although the diagnosis of vCJD may be impossible in the early stages, the combination of an affective or psychotic disorder with persistent pain, sensory symptoms, gait ataxia or dysarthria may at least raise the suspicion of the diagnosis of vCJD. This is particularly the case in younger patients in view of the age distribution of vCJD.

The variation in clinical presentation is reflected in the range of speciality at initial referral. Of the first 100 cases 38% of cases were first seen by a psychiatrist, 30% by a neurologist, 21% by a physician in general medicine and 11% by other specialists. Accurate diagnosis of vCJD was impossible prior to the description of the clinical phenotype in 1996. However, the range of initial diagnoses in the first 100 cases, the great majority of which presented after the publication of the first series of cases of vCJD, emphasizes the difficulty of early diagnosis (Table 1).

The later stages of vCJD are clinically similar to sCJD. After a mean of 6 months the majority of cases develop a progressive cerebellar syndrome with predominant ataxia of gait, although limb ataxia and dysarthria may also occur. At around the same time there is clear evidence of progressive cognitive impairment, leading to frank dementia. Involuntary movements are prominent and include chorea, dystonia and myoclonus and in some cases initial chorea evolves into a myoclonic movement disorder. Other features may develop, including dysphasia, pyramidal signs and rigidity of the limbs. The neurological deficits are progressive and cumulative. Terminally ill patients become totally dependent often leading to a state of akinetic mutism, although in some cases some limited response may be apparent even in the late stages. Death is usually due to intercurrent infection.

Table 1 Initial diagnoses in 100 cases of vCJD

Diagnostic cluster	*n*
Psychiatric	
Depression	42
Paranoid psychosis	3
Post-natal depression	4
Post-viral fatigue	4
Anxiety	2
Other neurotic	5
Neurological	
CJD	6
Encephalopathy	3
Progressive disorder	7
Multiple sclerosis	7
Cerebral lesion	2
Cerebellar lesion	3
Other organic	4
Mixed psychiatric and neurological	
Mixed	2
Orthopaedic	
Orthopaedic	2
No diagnosis	
None	4

4 Diagnosis and Investigation

The diagnosis of vCJD is likely to become suspected with the evolution and progression of frank neurological deficits. There is a range of potential differential diagnoses, including structural lesions, inflammatory disease and metabolic disorders, for example Wilson's disease. Appropriate investigation is essential to exclude alternative diagnoses and this requires referral to a specialist neurological centre.

A young age at death is also a feature in some forms of iatrogenic and familial CJD. It is essential to exclude past exposure to human pituitary growth hormone before considering the diagnosis of vCJD. Sequencing of the prion protein gene (*PRNP*) to identify mutations in the open reading frame may be necessary to distinguish in life between vCJD and genetic forms of human prion disease. Codon 129 of PRNP is a polymorphic region and is a recognized risk factor for sCJD with about 70% of cases having this genotype in comparison to about 40% of the general Caucasian population (Alperovitch et al. 1999). In vCJD all tested cases to date (111/130 in the UK and 121/140 worldwide) have been methionine homozygotes at codon 129 of PRNP. There is a possibility that bovine spongiform encephalopathy (BSE) infection in individuals with an alternative codon 129 genotype could occur in the future with a longer incubation period and possibly with a different clinical and/or pathological phenotype. To date there is no evidence of such cases.

Routine haematological and biochemical investigations are usually normal in all forms of CJD, except for abnormalities reflecting intercurrent illness. The electroencephalogram (EEG) is diagnostically useful in sCJD, showing triphasic periodic complexes in about 70% of cases (Bortone et al. 1994). In vCJD the EEG does not show these appearances, but may show non-specific generalized slowing. In some cases the EEG is normal and there may be a normal recording relatively late in the clinical course at a time when there are neurological and cognitive deficits. The cerebrospinal fluid (CSF) is usually normal in CJD, except for a modestly elevated protein level in some cases. The 14.3.3 CSF immunoassay is a marker of neuronal damage and, although a non-specific investigation, has been found to be a useful adjunct to diagnosis in sCJD (Zerr et al. 2000). In vCJD the test is positive in about 50% of cases and has a low specificity, because of the high positive rate in sCJD (Green et al. 2001).

In vCJD the most helpful investigation is magnetic resonance imaging (MRI) of the brain (Collie et al. 2001), which shows symmetrical high signal in the posterior thalamus on T2, PD or FLAIR images in about 90% of cases and this investigation has been included as a component of the diagnostic criteria for vCJD. The MRI scan shows abnormalities in the majority of cases of sCJD, but the distribution of high signal change, in the caudate and putamen, is distinct from that in vCJD (Finkenstaedt et al. 1996).

Brain biopsy has been used in the diagnosis of sCJD and vCJD. A positive result allows a definite diagnosis, but a negative result does not exclude the possibility of CJD because of the potential for sampling error. Brain biopsy has a significant risk of adverse effects and is usually reserved for cases in which there is a real possibility of an alternative treatable disorder such as cerebral vasculitis. Tonsil biopsy has been advocated as a diagnostic tool in vCJD, because, in contrast with other forms of human prion disease, systemic lymphoreticular tissue contains signifi-

Table 2 Differences between sporadic and variant CJD

	Sporadic	Variant
Mean age at death	66 years	29 years
Median duration of illness	4 months	14 months
Thalamic MRI high signal	Caudate/Putamen 60%	Pulvinar 90%
EEG	'Typical' 70%	'Typical' 0%

cant and identifiable levels of prion protein (Hill et al. 1999). Tonsil biopsy is an invasive procedure, which has risks, and may be most appropriate in those cases of suspect vCJD in which there is diagnostic uncertainty, for example cases with a negative MRI brain scan.

In cases of vCJD that die the most frequent differential diagnosis is sCJD. Table 2 summarizes the clinical and investigative differences between vCJD and sCJD.

5
Demographics of vCJD

The systematic classification of cases of vCJD depends on the formulation and validation of criteria for diagnosis. Partially validated diagnostic criteria for vCJD were first published in 2000 (Will et al. 2000) (Table 3). Cases can be classified as 'definite' with neuropathological confirmation, as 'probable' on the basis of clinical features and investigations or 'possible' based on clinical characteristics. The current sensitivity of a 'probable' diagnosis exceeds 80% and the specificity is 100%; there have been, as yet, no cases of 'probable' vCJD with an alternative final neuropathological diagnosis. 'Possible' cases are infrequent and are not included in reported figures or scientific analyses because of diagnostic uncertainty.

In the UK to date (February 2002) 94 definite and 36 probable cases have been identified, with six of these 'probable' cases still alive. Cases of vCJD have been confirmed in other countries (Table 4), but it is important to stress that cases are classified geographically by the country of normal residence at the time of disease onset. This does not necessarily imply the country in which exposure to BSE took place. For example, the cases of vCJD in the USA and Canada both had a history of extended residence in the UK during the 1980s when human dietary exposures to

Table 3 Diagnostic criteria for vCJD

I	A	Progressive neuropsychiatric disorder
	B	Duration of illness >6 months
	C	Routine investigations do not suggest an alternative diagnosis
	D	No history of potential iatrogenic exposure
	E	No evidence of a familial form of TSE
II	A	Early psychiatric symptoms[a]
	B	Persistent painful sensory symptoms[b]
	C	Ataxia
	D	Myoclonus or chorea or dystonia
	E	Dementia
III	A	EEG does not show the typical appearance of sporadic CJD[c] (or no EEG performed)
	B	Bilateral pulvinar high signal on MRI scan
IV	A	Positive tonsil biopsy[d]
DEFINITE:	I A and neuropathological confirmation of vCJD[e]	
PROBABLE:	I and 4/5 of II and III A and III B or I and IV A[d]	
POSSIBLE:	I and 4/5 of II and III A	

[a] Depression, anxiety, apathy, withdrawal, delusions.

[b] This includes both frank pain and/or dysaesthesia.

[c] Generalized triphasic periodic complexes at approximately one per second.

[d] Tonsil biopsy is not recommended routinely, nor in cases with EEG appearances typical of sporadic CJD, but may be useful in suspect cases in which the clinical features are compatible with vCJD and MRI does not show bilateral pulvinar high signal.

[e] Spongiform change and extensive PrP deposition with florid plaques, throughout the cerebrum and cerebellum.

Table 4 Cases of vCJD in the UK and elsewhere

Country	Number of cases	Year first identified
UK	130	1995
France	6	1996
Republic of Ireland	1	1999
Italy	1	2002
USA	1	2002
Canada	1	2002

BSE were likely to have been extensive. Although attributed to the USA and Canada, these cases are more likely to have been infected by BSE in the UK than in the country of residence at the time of diagnosis.

6
Conclusion

The clinical features of vCJD have been defined and diagnostic criteria formulated. Accurate diagnosis in life is possible in the majority of cases, but a firm diagnosis still depends on neuropathological verification. Early diagnosis of vCJD is important for the patient and their relatives and may have implications for the timeliness of public health measures to reduce the risk of secondary iatrogenic transmission. There is an imperative to develop techniques to allow prompt diagnosis and this will become pressing if effective treatments are developed.

References

Alperovitch A, Zerr I, Pocchiari M, Mitrova E, de Pedro Cuesta J, Hegyi I, Collins S, Kretzschmar H, van Duijn C, Will RG (1999) Codon 129 prion protein genotype and sporadic Creutzfeldt-Jakob disease. Lancet 353:1673–1674

Bortone E, Bettoni L, Giorgi C, Terzano MG, Trabattoni GR, Mancia D (1994) Reliability of EEG in the diagnosis of Creutzfeldt-Jakob disease. Electroencephalography & Clinical Neurophysiology 90:323–330

Brown P, Cathala F, Castaigne P, Gajdusek DC (1986) Creutzfeldt-Jakob disease: clinical analysis of a consecutive series of 230 neuropathologically verified cases. Ann Neurol 20:597–602

Collie DA, Sellar RJ, Zeidler M, Colchester AFC, Knight R, Will RG (2001) MRI of Creutzfeldt-Jakob disease: imaging features and recommended MRI protocol. Clinical Radiology 56:726–739

Finkenstaedt M, Szudra A, Zerr I, Poser S, Hise JH, Stoebner JM, Weber T (1996) MR imaging of Creutzfeldt-Jakob disease. Radiology 199:793–798

Green AJE, Thompson EJ, Stewart GE, Zeidler M, Mackenzie JM, Macleod MA, Ironside JW, Will RG, Knight RSG (2001) Use of 14-3-3 and other brain-specific proteins in CSF in the diagnosis of variant Creutzfeldt-Jakob disease. JNNP 2001:744–748

Henry C, Lowman A, Will RG (2002) Creutzfeldt-Jakob disease in elderly people. Age and Ageing 31:7–10

Hill AF, Butterworth RJ, Joiner S, Jackson G, Rossor MN, Thomas DJ, Frosh A, Tolley N, Bell JE, Spencer M, King A, Al-Sarraj S, Ironside JW, Lantos PL, Collinge J (1999) Investigation of variant Creutzfeldt-Jakob disease and other human prion diseases with tonsil biopsy samples. Lancet 353:183–184

Ironside JW, Head MW, Bell JE, McCardle L, Will RG (2000) Laboratory diagnosis of variant Creutzfeldt-Jakob disease. Histopathology 37:1–9

McNaughton HK, Will RG (1997) Creutzfeldt-Jakob disease presenting acutely as stroke: an analysis of 30 cases. Neurological Infections and Epidemiology 2:19–24

Parchi P, Giese A, Capellari S, Brown P, Schulz-Schaeffer W, Windl O, Zerr I, Budka H, Kopp N, Piccardo P, Poser S, Rojiani A, Streichemberger N, Julien J, Vital C, Ghetti B, Gambetti P, Kretzschmar H (1999) Classification of sporadic Creutzfeldt-Jakob disease based on molecular and phenotypic analysis of 300 subjects. Ann Neurol 46:224–233

Spencer MD, Knight RSG, Will RG (2002) First hundred cases of variant Creutzfeldt-Jakob disease: retrospective case note review of early psychiatric and neurological features. BMJ 324:1479–1482

Verity CM, Nicoll A, Will RG, Devereux G, Stellitano L (2000) Variant Creutzfeldt-Jakob disease in UK children: a national surveillance study. Lancet 356:1224–1227

Will RG, Ironside JW, Zeidler M, Cousens SN, Estibeiro K, Alperovitch A, Poser S, Pocchiari M, Hofman A, Smith PG (1996) A new variant of Creutzfeldt-Jakob disease in the UK. Lancet 347:921–925

Will RG, Stewart G, Zeidler M, Macleod MA, Knight RSG (1999) Psychiatric features of new variant Creutzfeldt-Jakob disease. Psychiatric Bulletin 23:264–267

Will RG, Zeidler M, Stewart GE, Macleod MA, Ironside JW, Cousens SN, Mackenzie J, Estibeiro K, Green AJE, Knight RSG (2000) Diagnosis of new variant Creutzfeldt-Jakob disease. Ann Neurol 47:575–582

Zeidler M, Johnstone EC, Bamber RWK, Dickens CM, Fisher CJ, Francis AF, Goldbeck R, Higgo R, Johnson-Sabine EC, Lodge GJ, McGarry P, Mitchell S, Tarlo L, Turner M, Ryley P, Will RG (1997) New variant Creutzfeldt-Jakob disease: psychiatric features. Lancet 350:908–910

Zeidler M, Stewart GE, Barraclough CR, Bateman DE, Bates D, Burn DJ, Colchester AC, Durward W, Fletcher NA, Hawkins SA, Mackenzie JM, Will RG (1997) New variant Creutzfeldt-Jakob disease: neurological features and diagnostic tests. Lancet 350:903–907

Zerr I, Pocchiari M, Collins S, Brandel J-P, de Pedro Cuesta J, Knight RS, Bernheimer H, Cardone F, Delasnerie-Laupretre N, Cuadrado Corrales N, Ladogana A, Bodemer M, Fletcher A, Awan T, Ruiz Bremon A, Budka H, Laplanche J-L, Will RG, Poser S (2000) Analysis of EEG and CSF 14-3-3 proteins as aids to the diagnosis of Creutzfeldt-Jakob disease. Neurology 55(6): 811–815

Neuropathology and Molecular Biology of Variant Creutzfeldt–Jakob Disease

J. W. Ironside · M. W. Head

National Creutzfeldt-Jakob Disease Surveillance Unit, Department of Pathology, Western General Hospital, University of Edinburgh, Edinburgh, EH4 2XU, UK
E-mail: j.w.ironside@ed.ac.uk

Abstract The neuropathological features of human prion diseases are spongiform change, neuronal loss, astrocytic proliferation and the accumulation of PrP^{Sc}, the abnormal isoform of prion protein (PrP). The pattern of brain involvement is remarkably variable and is substantially influenced by the host PrP genotype and PrP^{Sc} isotype. Variant Creutzfeldt–Jakob disease (vCJD) is a novel human prion disease which results from exposure to the bovine spongiform encephalopathy (BSE) agent. The neuropathology of vCJD shows consistent characteristics, with abundant florid and cluster plaques in the cerebrum and cerebellum, and widespread accumulation of PrP^{res} on immunocytochemistry. These features are distinct from all other types of human prion disease. Spongiform change is most marked in the basal ganglia, while the thalamus exhibits severe neuronal loss and gliosis in the posterior nuclei. These areas of thalamic pathology correlate with the areas of high signal seen in the thalamus on magnetic resonance imaging (MRI) examination of the brain. Western blot analysis of PrP^{Sc} in the brain in vCJD tissue shows a uniform isotype, with a glycoform ratio characterized by predominance of the diglycosylated band, distinct from sporadic CJD. PrP^{Sc} accumulation in vCJD is readily detectable outside the brain, in contrast with other forms of human prion disease, particularly in the lymphoid system and in parts of the peripheral nervous system. This has raised concern about the possible iatrogenic transmission of vCJD by contaminated surgical instruments, or blood. All cases of vCJD are methionine homozygotes at codon 129 of the prion protein gene (*PRNP*). Continued surveillance is required to investigate cases of vCJD in the UK and other countries where BSE has been reported, particularly as cases of 'human BSE' in individuals who are MV or VV at codon 129 of the PrP gene have not yet been identified. Histological, genetic and biochemical techniques are essential tools for the adequate diagnosis and investigation of human prion diseases.

1 Introduction

Surveillance of Creutzfeldt–Jakob disease (CJD) was re-established in 1990 in the UK in order to identify any changes in the incidence or characteristics of human prion diseases which might be the consequence of human exposure to the bovine spongiform encephalopathy (BSE) agent. There had been an earlier surveillance project in England and Wales (Will and Matthews 1984), but the new surveillance project covered the entire UK and was based in Edinburgh, in a unit with facilities for clinical, epidemiological and laboratory studies.

BSE was first reported by Wells et al. in 1987 in the UK, and subsequently occurred as an epidemic in UK cattle, particularly in the south of England. Early epidemiological studies identified meat and bonemeal animal feed as the likely source of the infection (Wilesmith et al. 1988); this was subsequently banned for use in cattle, but it is likely that exposure continued until at least the early 1990s (reviewed in the BSE Inquiry, 2000). It has been calculated that over one million cattle may have

Table 1 BSE-related diseases in other animal species in the UK

Species	Number of cases
Kudu	6
Gemsbok	1
Nyala	1
Oryx	2
Eland	6
Cheetah	5
Bison	1
Ankole cow	2
Puma	3
Tiger	3
Ocelot	3
Lion	3
Domestic cat	88

been infected by BSE, but the number of clinical cases is much lower (currently over 180,000 cases in the UK) (Anderson et al. 1996; DEFRA 2002). BSE is a spongiform encephalopathy which is characterized by vacuolation in the brainstem, with variable cerebral cortical pathology. The disease-associated form of the prion protein, PrP^{Sc}, and infectivity are readily detectable in the brainstem and in other brain regions, but there is no evidence of disseminated involvement of lymphoid tissues during the incubation period or in the clinical disease.

The consequences of potential human exposure to BSE were highlighted by the identification of novel prion diseases in antelopes in UK zoos, and in domestic and wild cats (Table 1) (DEFRA 2002). These animals presumably became infected with BSE by the oral route, and it is interesting to note that the neuropathology of the BSE-related illness in these species is quite distinct from the neuropathology of BSE in cattle. Experimental strain typing studies on cattle with BSE showed that a single strain of infectious agent was present in all cases (Bruce et al. 1994). This finding is in contrast with natural scrapie, where distinct strains have been identified in terms of their incubation period and neuropathology that results in strains of inbred mice following intracerebral inoculation. Similar studies on brain tissue from antelopes and cats with novel prion diseases demonstrated that the transmissible agent in these species exhibited identical strain qualities to the BSE agent, thereby confirming that transmission of BSE had occurred across several species barriers (Bruce et al. 1994).

The surveillance mechanism for CJD in the UK involves detailed clinical and epidemiological studies to identify potential risk factors for CJD, and laboratory studies to examine the neuropathological features of all cases of CJD identified in the UK. This is supplemented by analysis of the prion protein gene (*PRNP)*, and biochemical studies of PrP^{Sc} in the brain and other tissues. These studies were essential in the identification of variant CJD (vCJD) as a distinct entity within the spectrum of human prion diseases.

2
Human Prion Diseases

2.1
Sporadic CJD

Human prion diseases are a diverse group of fatal neurodegenerative disorders that occur in sporadic, familial and acquired forms (Table 2) (Prusiner 1993; Ironside 1998; Will et al. 1999). The commonest of these is the sporadic form of CJD (sCJD), which occurs as a worldwide disorder with a relatively uniform incidence of around one case per million of the population per annum (Will et al. 1999). Sporadic CJD accounts for around 85% of all human prion diseases. It exhibits considerable diversity in both its clinical and pathological features, which are substantially influenced by the PrP^{Sc} isotype (see below) and the naturally occurring polymorphism at codon 129 in the *PRNP* gene (Table 3). Large multicentre studies of sCJD have proposed a subclassification which is based on PrP^{Sc} isotype and *PRNP* codon 129 genotype (Parchi et al. 1996, 1999). It remains to be seen whether this classification will account

Table 2 Classification of human prion diseases

Idiopathic		Sporadic CJD Sporadic fatal insomnia
Familial		Familial CJD Fatal familial insomnia GSS (classical and variants)
Acquired	Human–human	Kuru Iatrogenic CJD
	Bovine–human	Variant CJD

Table 3 *PRNP* codon 129 genotypes in prion diseases

Genotype frequency (%)	MM	MV	VV
Normal Caucasian population	38	51	11
Sporadic CJD	71	15	14
Variant CJD	100	0	0

for all of the phenotypic diversity in sCJD, but it is nevertheless a useful framework for investigation and diagnosis.

Most cases of sCJD occur in patients in the seventh decade of life, but a wide age range has been reported from teenagers to extreme old age. The disease usually presents as a rapidly progressive dementia, with myoclonus, visual abnormalities and pyramidal or extrapyramidal signs (Brown et al. 1994; Will et al. 1999). The electroencephalogram (EEG) is abnormal in around 65% of patents, with triphasic periodic synchronous discharges. Magnetic resonance imaging (MRI) scans in sCJD exhibit areas of high intensity in the basal ganglia in most patients, and levels of the brain-derived protein 14.3.3 are increased in the cerebrospinal fluid (CSF) in around 90% of cases. Death usually occurs 4–6 months after the onset of disease, usually from bronchopneumonia.

Neuropathologically, most cases of sCJD exhibit spongiform change in the cerebral cortex in a widespread distribution, with variable neuronal loss and gliosis. Amyloid plaques are present in only a small subset of sCJD cases (MV2 subtype), usually in the cerebellum (Parchi et al. 1999). PrP^{Sc} accumulation is widespread in the grey matter, but with a variety of patterns which have been described as synaptic, perivacuolar, plaque-like and neuronal. PrP^{Sc} has not been identified in lymphoid tissues in sCJD, and there are inconsistent reports of PrP^{Sc} detection in the peripheral nervous system, e.g. dorsal root ganglia (Hainfellner et al. 1999).

2.2
Familial Prion Diseases

Familial forms of human prion disease account for around 10%–15% of all cases (Prusiner et al. 1993; Parchi et al. 1998c). These occur as autosomal dominant disorders with a high degree of penetrance, with an onset usually in midlife. All are associated with pathogenic mutations or insertions in the open reading frame of the *PRNP* gene, and there is a wide spectrum of phenotypic diversity both between cases with different mutations and within families with the same mutation. The first of the pathogenic mutations to be identified was the codon P102L mutatation in the Gerstmann–Straussler–Scheinker syndrome (GSS) (Hsaio et al. 1989). Since then, an ever-increasing number of *PRNP* gene mutations have been identified (for a review see Parchi et al. 1998c).

The clinical features of inherited prion diseases are too complex to discuss in detail here, but four main phenotypes have been described: a sCJD-like phenotype, the GSS phenotype (with progressive cerebellar ataxia, other movement disorders and dementia relatively late in the disease), fatal familial insomnia (sleep disorder with autonomic dysfunction, variable ataxia and late dementia) and an intermediate phenotype which is usually associated with pathogenic insertions in the octapeptide repeat region of the *PRNP* gene (Parchi et al. 1998c). The pathological features are also diverse, and four similar groups can be identified, of which perhaps the most striking is the GSS-like group, where numerous large multicentric amyloid plaques are present in a wide distribution in the brain. Fatal familial insomnia is characterized by thalamic gliosis and neuronal loss, usually involving the anterior and medial thalamic nuclei. Both the clinical and pathological features of these diseases are also influenced by the codon 129 polymorphism in the *PRNP* gene; a good example of this effect is the relationship between fatal familial insomnia, where the D178N mutation is linked with methionine at codon 129 in the mutated allele. In contrast if valine is coupled with the mutated allele, this allotype results in a disease phenotype resembling sCJD (Gambetti et al. 1995).

2.3
Acquired Prion Diseases

The acquired human prion diseases include kuru, iatrogenic CJD and vCJD. Kuru was described in the 1950s (Gajdusek and Zigas 1957), and occurred as an epidemic in the Fore tribe of Papua New Guinea. Clinical features included prominent cerebellar ataxia with movement disorders and late dementia, and the pathology showed cerebellar atrophy usually accompanied by rounded amyloid plaques, known as kuru plaques. Iatrogenic transmission of CJD has occurred following the use of contaminated neurosurgical instruments and intracerebral electrodes, corneal

Table 4 Worldwide cases of iatrogenic CJD

Mode of infection	Number of cases	Incubation period (months)
Neurosurgical instruments:		
Neurosurgery	5	12–28
Stereotactic EEG	2	16–20
Tissue transplant:		
Corneal transplant	4	16–320
Dura mater transplant	120	18–216
Human pituitary hormone injection:		
Growth hormone	142	50–456
Gonadotrophin	5	144–192

and dura mater grafts, and in the recipients of human pituitary hormones (Table 4) (Will et al. 1999). As might be expected from such diverse sources of infection, there are few consistent clinical or pathological features in this group (for a review see Brown et al. 2000). However, it has been observed that the incubation periods following peripheral infection (e.g. in human pituitary hormone recipients) are prolonged, and can last for over 30 years.

3 Prion Biology

3.1 PrP Biochemistry

The prion protein (PrP) is a copper-binding glycoprotein of uncertain function, expressed in a wide variety of cell types, but especially abundantly in central nervous system (CNS) neurones, where it localizes to synapses. The pathological conversion of this normal form of the prion protein (PrP^C) to an abnormal disease-associated isoform (PrP^{Sc}) is thought to be *a*, if not *the*, central event in the pathogenesis of prion diseases. The biochemical properties of PrP^C and PrP^{Sc} contrast starkly: PrP^C is a monomeric protein with a globular C-terminal domain rich in α-helical structure. In contrast, PrP^{Sc} is rich in β-sheet, highly aggregated and contains a C-terminal core with a considerable resistance to proteolytic degradation in non-denaturing conditions. Unique differences in the primary structure or covalent modification of the two forms have yet to be described. Instead this conversion is thought to be conformational in nature. A number of models have been proposed to describe this process, but perhaps the leading candidate involves the 'seeded polymerization' of PrP^{C} by aggregated PrP^{Sc} in a two-step (binding and conversion) process. How this process might result in neurodegeneration is currently unknown but it quite possibly involves both the loss of an antioxidant property associated with PrP^{C} expression and the acquisition of cytotoxic properties by PrP^{Sc} (reviewed by Giese and Kretzschmar 2001). With only a few exceptions there is a close correlation between PrP^{Sc} and infectivity, as one would predict from the protein-only version of the prion hypothesis which identifies PrP^{Sc} itself as a novel *pro*teinasceous *in*fectious agent or *prion* (reviewed by Prusiner 1998). A particular problem for the prion hypothesis has been the well-documented existence of multiple sheep scrapie strains, as defined by incubation period and lesion profile on transmission to rodents. One possible solution to this problem became apparent when two differing rodent-adapted transmissible mink encephalopathy strains, termed *hyper* and *drowsy* (after their differing disease phenotypes) were found to have differently sized protease-resistant core fragments when analysed by Western blotting (Bessen and Marsh 1994). This has led to the proposition that PrP^{Sc} can exist in a series of self-propagating conformational variants and that these can in part confer the strain-like properties of prions.

3.2 Human PrP^{Sc} Isotypes

The phenotypes of the human prion diseases are diverse. In principle this may derive from a number of considerations including the interactions of agent strain, route of infection, and host factors, including mutations and polymorphisms of the host *PRNP* gene. The past few years has seen a considerable effort to identify aspects of biochemical variation in human PrP^{Sc} and to produce a classification scheme that might prove diagnostically useful. The biochemical properties investigated thus far include the following:

Primary Sequence. In most cases this is confined to mutations associated with inherited prion diseases and the polymorphic *PRNP* codon 129 which can encode either methionine or valine and is determined by DNA sequencing (Prusiner 1998).

Conformation. This has been addressed by the somewhat indirect method of determining the size of the protease-resistant core using Western blot analysis (Collinge et al. 1996; Parchi et al. 1996). For the sake of clarity the term PrP^{res} has been adopted to denote PrP^{Sc} defined operationally by its resistance to proteolytic degradation. Interestingly endogenously N-terminal truncated forms also occur in situ (Zanusso et al. 2001).

Metal Occupancy. PrP has N-terminal metal binding sites for copper and perhaps other divalent cations. Occupancy of these sites appears to affect conformation as treatment with appropriate chelating agents can in some cases change the protease-resistant core fragment size seen on Western analysis (Wadsworth et al. 1999a).

Glycosylation Site Occupancy. PrP has two asparagine-linked glycosylation sites of variable occupancy. The three glycoforms (di-, mono, and non-glycosylated) of PrP^{Sc} are resolved during Western analysis and following densitometric analysis can be expressed as a glycoform ratio (see for example Collinge et al. 1996).

Glycan Composition. While methods exist that determine the exact composition of the glycans, two-dimensional gel electrophoresis has recently started to be used to give an idea of the complexity of the glycans as a function of their charge heterogeneity (Pan et al. 2001; Zanusso et al. 2002).

The most consistently reported aspects of human PrP^{Sc} analysis are the protease-resistant core fragment size and the glycoform ratio of PrP^{res}. However this relatively simple form of analysis had proved surprisingly controversial and resulted in the formulation of two competing nomenclatures. In one, fragment size and glycoform ratio are considered independently and in the second they are considered together. The primary area of disagreement is in the numbers of possible protease-resistant core fragment sizes that can be resolved in different cases of sCJD and is therefore not directly pertinent to a discussion of vCJD. However the dual nomenclatures has resulted in confusion hence the most probable correlation of PrP^{res} isotypes is shown in Table 5.

Table 5 Probable correlation of PrP isotype nomenclatures

Collinge	Characteristics	Gambetti	Characteristics
Type 1	~22 kDa[a], M/M[b] sCJD brain, monoglycosylated forms predominate (Collinge et al. 1996)	1	21 kDa, M/M, M/V, V/V sCJD brain (Parchi et al. 1997, 2000)
Type 2	<type 1, M/M, M/V, V/V sCJD brain, monoglycosylated forms predominate (Collinge et al. 1996)	1	
Type 2⁻	<type 2 and >type 3, produced by experimental metal depletion of types 1 and 2 (Wadsworth et al. 1999a)	nd	
Type 3	2–3 kDa <type 2, M/V, V/V sCJD brain, monoglycosylated forms predominate (Collinge et al. 1996, Wadsworth et al. 1999b)	2A	19 kDa, M/M, M/V, V/V sCJD brain. Monoglycosylated forms predominate (Parchi et al. 1997, 2000)
Type 4	>type 3, M/M, vCJD brain, diglycosylated forms predominate (Collinge et al. 1996)	2B	19 kDa, M/M vCJD brain. Diglycosylated forms predominate (Parchi et al. 1997, 2000 Ironside et al. 2000)
Type 4t	vCJD tonsil with further accentuation of type 4 glycoform ratio (Hill et al. 1999)	nd	
Type 5	vCJD transmitted to humanized V/V mice. Type 2 mobility and type 4 glycoform ratio (Hill et al 1997)	nd	

[a] The relative mobility of the non-glycosylated PrP^{res} is given in kDa.

[b] The *PRNP* polymorphic codon 129 is indicated as M (methionine) and V (valine).

[c] nd, Not determined.

4
Variant CJD

Variant CJD was identified as a novel human prion disease in the UK in 1996 in a series of 10 patients (Will et al. 1996). One hundred and forty three cases of vCJD have so far been identified in the UK, with six cases in France and one each in Canada, Ireland, Italy, the USA and Canada. All cases of vCJD have occurred in individuals who were homozygous for methionine at codon 129 in the *PRNP*. The clinical features of vCJD have remained relatively consistent since the first description, and comprise pyschiatric symptoms at onset, such as anxiety, personality change and a loss of concentration (Spencer et al. 2002). These are accompanied or followed shortly by sensory abnormalities such as paraesthesiae or dysaesthesiae, which often involve the face, back or lower limb. Cerebellar ataxia is a prominent feature and is usually accompanied by movement disorders, such as myoclonus or chorea (Will et al. 2000). The final stages of the illness include cognitive impairment and dementia, with a terminal akinetic mute state. The EEG in vCJD is abnormal, but does not show the periodic synchronous discharges seen in sCJD. Levels of 14.3.3 in the CSF are increased in only around 50% of patients, so this is not a diagnostically useful investigation (Green et al. 2001). In contrast, MRI scans show a symmetrical area of high signal in the posterior thalamus, often extending into the dorsmedial nucleus. This abnormality has a high level of sensitivity and specificity, making it an excellent non-invasive test in patients with suspected vCJD (Zeidler et al. 2000).

4.1 Neuropathology of vCJD

Since 1990, central neuropathological review has been performed in the CJD Surveillance Unit for all cases of CJD diagnosed in the UK. Tissues from autopsy cases are fixed in formalin for a minimum of 3 weeks. The brains are sampled to include the frontal, parietal, temporal and occipital cortex, the hippocampus, hypothalamus, thalamus, basal ganglia, midbrain, pons, medulla and spinal cord. Other organs are examined histologically if appropriate permission was obtained and material was available. Sections are cut at 5 μm and stained by conventional neuropathological techniques and by immunocytochemistry for PrP^{res} using a range of monoclonal antibodies in a standardized validated technique (Bell et al. 1995).

4.1.1 CNS Histology

The principal neuropathological features of vCJD are summarized in Table 6 (Ironside et al. 2000a). In most cases, the brain weight after fixation is within the normal range in relation to the age of the patient, but in cases with a lengthy clinical history (>18 months) there is evidence of cerebral cortical and cerebellar atrophy (particularly involving the cerebellar vermis). In the single case of vCJD identified so far in the geriatric population, the neuropathologial features were very similar to those occurring in the brains of younger patients (Lorains et al. 2001). On histology, florid plaques are readily identified on haematoxylin and eosin stains in all cases in the cerebral cortex (particularly the occipital cortex) and the cerebellar cortex. These comprise an eosinophilic central core with radiating fibrils in the periphery of the plaque, which are surrounded by a rim of spongiform change in an otherwise intact neuropil (Will et al. 1996) (Fig. 1). Florid plaques occur in all layers of the cerebral cortex, often in an apparently random focal distribution. In the cerebellum, florid plaques are most easily identified in the molecular layer (Fig. 1), occasionally projecting into the subpial space, and are also present within in the granular layer. Plaques are often present close to blood vessels, but amyloid angiopathy is absent.

Spongiform change within the cerebral cortex is usually most prominent in the occipital and inferior frontal cortex; the hippocampus exhibits little spongiform change. Spongiform change is a prominent fea-

Table 6 Diagnostic neuropathological features of vCJD

Cerebral and cerebellar cortex:	Multiple florid plaques in H&E sections Numerous small cluster plaques in PrP stained sections Amorphous pericellular and perivascular PrP accumulation
Caudate nucleus and putamen:	Severe spongiform change Perineuronal and axonal PrP accumulation
Posterior thalamic nuclei and midbrain:	Marked astrocytosis and neuronal loss
Brainstem and spinal cord:	Reticular and perineuronal PrP accumulation in grey matter

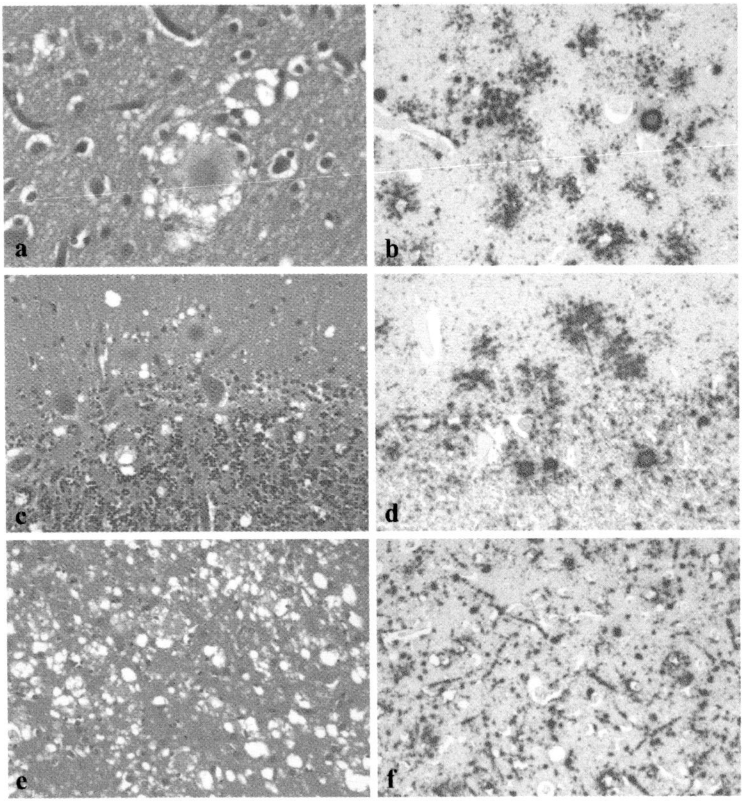

Fig. 1 **a** The cerebral cortex in vCJD contains florid plaques (*centre*) composed of a fibrillary amyloid core surrounded by spongiform change. H&E. **b** Immunocytochemistry for PrP^{res} in the cerebral cortex in vCJD shows intense staining of the florid plaques (*brown*), with widespread amorphous PrP^{res} deposits which are not visible on routine stains. KG9 monoclonal antibody. **c** The cerebellar cortex in vCJD contains amyloid plaques (*centre*), but there is less spongiform change than in the cerebral cortex. H&E. **d** Immunocytchemistry for PrP^{res} in the cerebellar cortex in vCJD shows a similar pattern of deposition to the cerebral cortex. KG9 monoclonal antibody. **e** The caudate nucleus in vCJD exhibits severe and widespread spongiform change. H&E. **f** Immunocytochemistry for PrP^{res} in the caudate nucleus in vCJD shows perineuronal and linear periaxonal positivity, in contrast with the cerebral and cerebellar cortex. KG9 monoclonal antibody

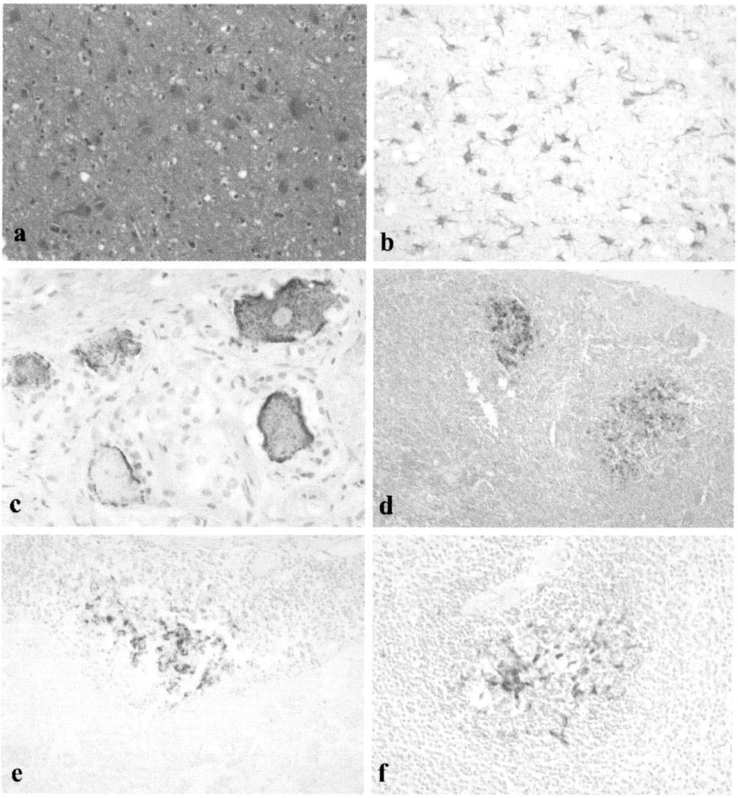

Fig. 2 **a** The posterior thalamus in vCJD shows mild spongiform change, with few plaques, but severe neuronal loss is evident. H&E. **b** Immunocytochemistry for glial fibrillary acidic protein in the thalamus in vCJD confirms the presence of a marked astroctytic reaction in areas of neuronal loss. **c** Positive staining for PrP^{res} in the ganglion and satellite cells within a spinal dorsal root ganglion in vCJD. KG9 monoclonal antibody. **d–f** Immunocytochemistry for PrP^{res} labels follicular dendritic cells within the germinal centres of lymphoid tissues throughout the body in vCJD: **d**, tonsil; **e**, appendix; **f**, spleen. 3F4 monoclonal antibody

ture in the cerebellar hemispheres and vermis. Confluent spongiform change is rarely present in the cerebral cortex, in comparison to cases of sCJD. However, extensive spongiform change is a constant feature in the caudate nucleus and putamen (Fig. 1). Similar changes are present in the anterior thalamus, but focal spongiform change is seldom present in the posterior thalamic nuclei. In these regions, the neuropathology is dominated by severe neuronal loss and marked astrocytosis (Fig. 2). The distribution of astrocytosis does not relate to the presence of amyloid plaques, but is closely associated with neuronal loss (Zeidler et al. 2000). Spongiform change is most evident in the paraventricular nuclei of the hypothalamus, which also contains occasional amyloid plaques. In the brainstem, spongiform change is present in the periaqueductal grey matter in the midbrain and in the pontine nuclei, but is not present in either the medulla or spinal cord. In the midbrain, severe neuronal loss and astrocytosis occurs in the colliculi and the periaqueductal grey matter.

4.1.2
PrP Immunocytochemistry in CNS

Immunocytochemistry for PrP^{res} shows intense staining of the florid plaques in all areas of the cerebral and cerebellar cortex (Fig. 1). This technique also reveals numerous smaller plaques which are not visible on routinely stained sections, often arranged in irregular clusters (Fig. 1). These 'cluster plaques' are present in all cases. Furthermore, there is a widespread deposition of PrP^{res} in a pericellular (neuronal and astrocytic) distribution in the cerebral and cerebellar cortex. These unusual deposits can also be visualized on Gallyas silver impregnation. Quantitative assessment has confirmed the marked PrP^{res} accumulation in the occipital cortex and the cerebellar cortex, in comparison with other cortical regions, basal ganglia and thalamus (Ironside et al. 1996). In the hippocampus, the cornu ammonis showed little PrP^{res} deposition, but there was a dense accumulation in the dentate fascia and the subiculum.

PrP^{res} accumulation in the basal ganglia and thalamus occurs in a linear and perineuronal pattern (Fig. 1). A reticular pattern of immunoreactivity for PrP^{res} is also detected in the thalamus and hypothalamus, with occasional PrP^{res} plaques. Synaptic and neuronal PrP^{res} positivity is present in the brainstem and spinal cord (particularly in the substantia gelatinosa). PrP^{res} immunoreactivity is not detected in the meninges (dura mater or arachnoid mater).

4.1.3 PrP Immunocytochemistry in Non-CNS Tissues

In the peripheral nervous system, PrP^{res} has been identified in the ganglion and satellite cells in dorsal root ganglia (Fig. 2) and trigeminal ganglia, and within the retina and optic nerve. Other peripheral nerves give a negative staining reaction for PrP^{res}. Positive staining for PrP^{res} occurs in follicular dendritic cells within germinal centres of the tonsil, spleen and lymph nodes (Fig. 2) (Hill et al. 1999). Positive staining for PrP^{res} was also identified in cells that correspond to follicular dendritic cells in lymphoid follicles in the wall of the appendix (Fig. 2), and in Peyer's patches in the ileum. In one patient who died from variant CJD, PrP^{res} was identified within lymphoid tissue in the appendix which had been removed 8 months prior to the onset of neurological disease (Hilton et al. 1999). As a result of this, a large-scale retrospective study to detect PrP^{res} in surgically resected appendices has commenced, in order to attempt to refine the number of individuals within the UK who might be incubating vCJD (Ironside et al. 2000b). A recent study using Western blotting to detect PrP^{res} in vCJD reported positive findings in the adrenal gland and rectum in addition to lymphoid tissues (thymus, spleen tonsil and lymph nodes) (Wadsworth et al. 2001). PrP^{res} immunocytochemistry in the other main organs (including the heart, lung, muscle, liver, kidney, bladder, testes, and pelvic organs and skin) are all negative for PrP^{res} (Ironside et al. 2000a). Infectivity in the tonsil and spleen in vCJD has been confirmed by experimental transmission in mice (Bruce et al. 2001), with levels of infectivity in the spleen around 2 logs lower than in the brain. Interestingly, attempts to demonstrate infectivity in the buffy coat fraction of blood were negative in this study.

4.2 PrP^{res} in vCJD

On the basis of the first ten cases of vCJD, Collinge and co-workers identified a characteristic glycoform ratio as the distinguishing feature of vCJD brain PrP^{res} (Collinge et al. 1996). This isotype had a characteristic predominance of the diglycosylated form that differentiated it from iatrogenic and sCJD. This glycotype was also found in BSE and in natural and experimental BSE transmission suggesting that it is the 'glycoform signature' of the BSE agent (Collinge et al. 1996; Hill et al. 1997).

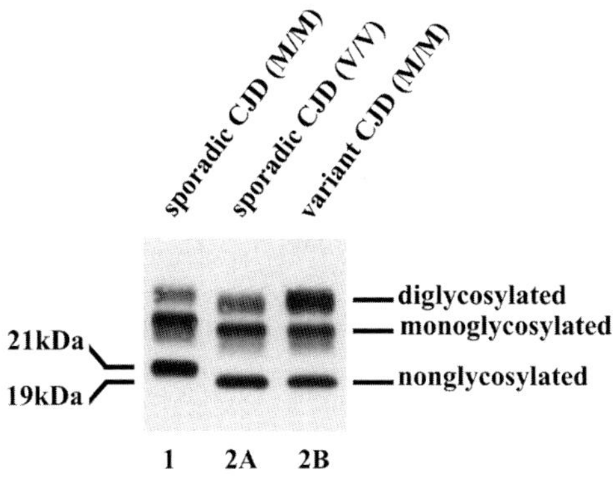

Fig. 3 Western blot analysis of frontal cortex from sporadic and variant CJD brain. The three glycoforms (di-, mono-, and non-glycosylated) of PrP^{res} are marked. The two most frequently occurring forms of sCJD, in methionine (*M/M*) and valine (*V/V*) homozygotes have non-glycosylated PrP^{res} of either 21 kDa (*type 1*) or 19 kDa, (*type 2*) respectively. PrP^{res} in vCJD has a type 2 mobility (19 kDa) but it is distinguished from the type 2 sporadic CJD isotype (*type 2A*) by the predominance of the diglycosylated form (*type 2B*)

Subsequent studies have shown this glycotype is a consistent feature of vCJD brain, present in all cases of vCJD (*n*=57) thus far analysed by the National CJD Surveillance Unit and unaffected by part of the brain assayed. A typical Western blot analysis of PrP^{res} from variant and sporadic CJD brain is shown in Fig. 3. The non-glycosylated protease-resistant PrP fragment from vCJD brain is approximately 19 kDa and N-terminal sequencing shows that the consensus N terminus is at serine 97, similar to type 2 PrP^{res} found in sporadic, iatrogenic and familial CJD (Parchi et al. 2000). This suggests that the conformation of vCJD PrP^{res}, as determined by current assays, is not unique but rather that the BSE agent consistently results in the accumulation of PrP^{res} with a higher glycosylation site occupancy than that of other forms of CJD. A mechanism where by a host controlled process (PrP^{C} glycosylation) can result in the accumulation of an agent-defined PrP^{Sc} glycotype is yet to be established (see Somerville 1999 for a discussion of the possibilities). Although characteristic, this glycoform ratio is not unique to vCJD: glycoform ratios similar to those characteristic of vCJD are also seen in some inherited (Parchi et al. 1998a, 1998b; Hainfellner et al. 1999) and sporadic (Head et al. 2001) cases of human prion disease.

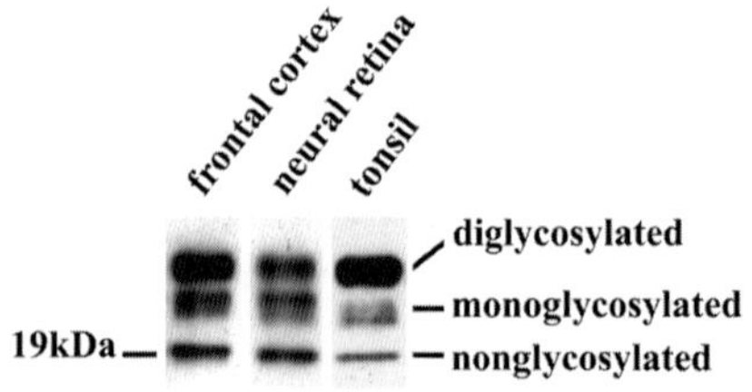

Fig. 4 Western blot analysis of PrP^{res} from vCJD brain (frontal cortex), eye (neural retina), and lymphoreticular system (tonsil). The three glycoforms of PrP^{res} (di-, mono-, and non-glycosylated) are marked, as is the relative mobility of the non-glycosylated band (19 kDa). Note that the relative proportions of the three glycoforms vary between samples. The diglycosylated form predominates in frontal cortex and its relative abundance is higher in tonsil and lower in retina

The complexity and composition of glycans themselves on PrP^{res} in different forms of sCJD has recently been proposed to be a marker of agent strain (Pan et al. 2001). Similar studies have yet to be reported for vCJD tissues.

While the human brain vCJD glycoform signature is diagnostically useful, it does however appear to be modulated by the cell types within the tissue of accumulation (see Fig. 4). PrP^{res} accumulated in the lymphoreticular system in variant CJD has an even more accentuated glycoform ratio than that found in corresponding brain samples (Hill et al. 1999). In contrast, the PrP^{res} found in the retina is not characterized by diglycosylated isoforms, underlining the role of host factors in determining PrP^{res} isotype (Head et al. 2003). The biochemical features of PrP^{res} in variant CJD are summarized in Table 7.

Table 7 Biochemical features of variant CJD PrP^{res}

Brain
Non-glycosylated PrP^{res} fragment 19 kDa on Western blot
Majority N terminus at serine 97 after proteinase K treatment
Di-glycosylated form predominates
Uniform isotype throughout the brain
Methionine at position 129
Peripheral organs
Present in the lymphoreticular system at levels lower than those found in brain
Glycoform ratio further accentuated towards diglycosylated forms in lymphoreticular system
Glycoform signature reduced in the retina

5 Relationship of vCJD to BSE

The evidence for a causal relationship between vCJD and BSE was first proposed on the basis of epidemiological evidence by Will et al. in 1996. The ensuing epidemiological evidence has continued to support this proposal, along with the results from several independent laboratory studies. Experimental transmission of BSE to three macaques revealed neuropathological changes which were remarkably similar to those of vCJD (Lasmezas et al. 1996). This observation has been confirmed in subsequent larger studies (Lasmezas et al. 2001). The biochemical similarities (in terms of glycoform ratios) of the PrP^{res} in the brain in vCJD to that in the brains of cattle with BSE also supported this theory, along with the observation that similar glycoform ratios were observed in the brains of several other species infected with BSE (Collinge et al. 1996). This evidence of a causal link between BSE and vCJD has been considerably strengthened by the results of experimental strain typing studies in inbred mice (Bruce et al. 1997), and subsequent transmission studies of BSE and vCJD in bovine transgenic mice (Scott et al. 1999). In the latter study, it was interesting to note that amyloid plaque formation was a characteristic of the neuropathology of vCJD and BSE transmission into the brains of bovine transgenic mice. Since it is likely that exposure to BSE has occurred in all prion protein genotypes in the population, the development of criteria to detect a BSE-related disease in individuals who are heterozygotes or valine homozygotes at codon 129 of the *PRNP* gene is of considerable importance. This question may only be resolved by experimental transmission studies into both inbred and transgenic mice.

6 Diagnostic and Clinicopathological Considerations

The World Health Organization has indicated that neuropathological examination is mandatory for the confirmation of a diagnosis of vCJD (WHO 1996). Neuropathological and biochemical studies on vCJD allow a diagnostic distinction from other forms of CJD; the relative uniformity of the neuropathology of vCJD is in contrast with that of sCJD (the main differential pathological diagnosis). This feature is also consistent with the effects of the BSE in patients with a uniform *PRNP* genotype (methionine homozygotes). In kuru (another acquired human prion disease which was probably also transmitted by the oral route), codon 129 *PRNP* genotype did not have an influence on the neuropathological or clinical features in one series (McLean et al. 1997), although in another study (without detailed pathological review) it was suggested that amyloid plaques were more numerous (but not exclusively present) in individuals who were valine homozygotes or heterozygotes at codon 129 (Cervenekova et al. 1999).

The florid plaque is a neuropathological hallmark of vCJD; these structures are present in a widespread distribution in the grey matter of cerebral and cerebellar cortex. Florid plaques have not been reported in cases of sCJD, and there seems to be no major overlap of the neuropathology of vCJD with any of the subtypes of sCJD identified so far. The neuropathological diagnostic criteria proposed here for vCJD allow a clear distinction from other human prion disease (Table 6). However, florid plaques are not unique to vCJD, having been first described in transmissions of Icelandic scrapie to mice (Fraser et al. 1979). They have also been reported in chronic wasting disease, particularly in white-tailed deer (Williams et al. 1993). Florid plaques have been reported recently in occasional cases of iatrogenic CJD in dura mater graft recipients in Japan (Shimizu et al. 1999); however, none of these cases exhibited any of the other neuropathological features of vCJD. The rounded amyloid plaques occurring in sCJD (MV2 subtype) and kuru can be distinguished easily from florid plaques by their restricted anatomical distribution, their compact morphology and by the absence of a corona of spongiform change in the surrounding neuropil (Parchi et al. 1999, Ironside et al. 2000a).

The distribution of thalamic neuronal loss and gliosis in vCJD is anatomically distinct from fatal familial insomnia (Gambetti et al. 1995) and sporadic fatal insomnia (Mastrianni et al. 1999). This is of particular interest in relation to the sensory abnormalities in patients with vCJD, which may represent thalamic pain. Further studies are required to assess the pathology in the posterior thalamic nuclei in vCJD, and to relate these changes to the sensory abnormalities occurring in this disorder.

The widespread distribution of PrP^{res} in vCJD is another characteristic feature which clearly distinguishes it from other human prion diseases. The involvement of the lymphoid and peripheral nervous systems in vCJD is similar to that occurring in experimental models of prion disease following oral exposure to the transmissible agent (Beekes and

McBride 2000). Thus, oral exposure to BSE (currently the most likely route of exposure) may be reflected in the peripheral pathogenesis of the vCJD as identified by PrP^{res} accumulation. If so, it is interesting to note that no evidence of PrP^{res} accumulation has been identified in other acquired forms of human prion disease, particularly in iatrogenic forms of CJD occurring after exposure to contaminated growth hormone (Hill et al. 1999). In these patients, injection of the hormone was given by the intramuscular or subdermal route, but so far no evidence of lymphoid tissue accumulation of PrP^{res} has been observed in such cases.

The analysis of PrP^{res} in the brain is important in discriminating between CJD and all other dementias. Our studies confirm that the PrP^{res} glycoform ratio can help distinguish between vCJD and sCJD (Ironside et al. 2000a), but a glycoform pattern similar to that of vCJD has been identified in both fatal familial insomnia and GSS (Parchi et al. 1995, 1999; Telling et al. 1996). In practical terms, this may not be such an important diagnostic issue as both fatal familial insomnia and GSS can be diagnosed with *PRNP* genotype analysis. Since no cases of vCJD have yet been identified in patients with codon 129 *PRNP* genotypes other than methionine homozygotes, it is not possible to comment on the likely pathological, biochemical or clinical features (if indeed this ever occurs). Experimental transmission studies using transgenic mice have suggested that the currently known glycoform ratio of vCJD will be preserved following BSE infection in humans with valine homozygous or heterozygous *PRNP* genotypes (Collinge et al. 1995; Hill et al. 1997). This will certainly aid diagnosis, but as discussed above, unequivocal identification of such cases will probably depend on the results of experimental strain typing in mice.

The widespread accumulation of PrP in lymphoid tissues in vCJD has given rise to concerns that the disease might be transmitted accidentally by surgical instruments which have come into contact with infected lymphoid tissues. Since conventional forms of hospital decontamination cannot remove all traces of prion infectivity, this theoretical risk is currently under evaluation in the UK, along with steps for risk reduction and risk management. Concerns that blood might contain infectivity in vCJD (perhaps associated with circulating lymphoid cells) has led to the introduction of universal leucodepletion in the UK, and sourcing of plasma products outside the UK (Turner 2001). These actions have been taken on the basis of a theoretical risk, and until a satisfactory screening test is developed to detect asymptomatic individuals who are incubating vCJD, these precautionary measures will probably have to stay in place in order to help protect public health. The concerns over possible infectivity in blood in vCJD have been reinforced by the recent demonstration of experimental BSE transmission by blood transfusion in a sheep model (Hunter et al. 2002).

7 Conclusions

Variant CJD is a novel human prion disease with distinctive pathological and biochemical features. All cases investigated so far are methionine homozygotes at codon 129 in the *PRNP*. It results from human exposure to the BSE agent, although the precise mechanisms of exposure are uncertain. The widespread distribution of PrP^{res} in lymphoid tissues in vCJD is different from all other human prion diseases, and has raised concerns about potential iatrogenic transmission by contaminated surgical instruments and blood. The diagnosis of vCJD needs an integrated approach using a combination of neuropathology, *PRNP* sequencing and PrP^{res} analysis (Ironside et al. 2000). On this basis, there is no evidence to suggest that BSE infection has yet been identified in patients with valine homozygous or heterozygous *PRNP* genotypes. Continuing surveillance with detailed laboratory investigation of all suspect CJD cases is required in all countries with reported cases of BSE.

Acknowledgements. We thank all the neuropathologists in the United Kingdom, and our colleagues Prof. R.G. Will and Dr R. Knight, Mr M. Bishop and Dr T. Bunn. We thank Mrs L. McCardle and her staff for technical assistance and Ms D. Ritchie for expert photographic assistance. The CJD Surveillance Unit is funded by the Department of Health and the Scottish Executive Department of Health. This work was supported by EU Bio4-CT98-6046.

References

Anderson RM, Donnelly CA, Ferguson NM, Woolhouse ME, Watt CJ, Udy HJ, MaWhinney S, Dunstan SP, Southwood TR, Wilesmith JW, Ryan JB, Hoinville LJ, Hillerton JE, Austin AR, Wells GA (1996) Transmission dynamics and epidemiology of BSE in British cattle. Nature 382:779–88

Bell JE, Gentleman SM, Ironside JW, McCardle L, Lantos PL, Dey L, Lowe J, Fergusson J, Luthert P, McQuaid S (1997) Prion protein immunocytochemistry—UK five centre consensus report. Neuropathol Appl Neurobiol 23:26–35

Beekes M, McBride P (2000) Early accumulation of pathological PrP in the enteric nervous system and gut-associated lymphoid tissue of hamsters orally fed with scrapie. Neurosci Lett 278:81–184

Bessen RA, Marsh RF (1994) Distinct PrP properties suggest the molecular basis of strain variation in transmissible mink encephalopathy. J Virol 68:7859–7868

Brown P, Gibbs CJ Jr, Rodgers-Johnson P, Asher DM, Sulima MP, Bacote A, Goldfarb LG, Gajdusek DC (1994) Human spongiform encephalopathy: the National Institutes of Health series of 300 cases of experimentally transmitted disease. Ann Neurol 35:513–529

Brown P, Preece P, Brandel JP, Sato T, McShane L, Zerr I, Fletcher A, Will RG, Pocchiari M, Cashman NR, d'Aignaux JH, Cervenakova L, Fradkin J, Schonberger LB, Collins SJ (2000) Iatrogenic Creutzfeldt-Jakob disease at the millennium. Neurology 55:1075–1081

Bruce ME, Will RG, Ironside JW, McConnell I, Drummond D, Suttie A, McCardle L, Chree A, Hope J, Birkett C, Cousens S, Fraser H, Bostock C (1997) Transmissions to mice indicate that "new variant" CJD is caused by the BSE agent. Nature 389:498–501

Bruce ME, Chree A, McConnell I, Foster J, Pearson G, Fraser H (1994) Transmission of bovine spongiform encephalopathy and scrapie to mice: strain variation and the species barrier. Philos Trans R Soc Lond B Biol Sci 343:405–411

Bruce ME, McConnell I, Will RG, Ironside JW (2001) Detection of variant Creutzfeldt-Jakob disease infectivity in extraneural tissues. Lancet 358:208–209

Cervenakova L, Goldfarb LG, Garruto R, Lee H-S, Gajdusek DC, Brown P (1999) Phenotype-genotype studies in kuru: implications for new variant Creutzfeldt-Jakob disease. Proc Natl Acad Sci USA 95:13239–13241

Collinge J, Palmer MS, Sidle KC, Hill AF, Gowkland I, Meads J, Asante E, Bradley R, Doey LJ, Lantos P (1995) Unaltered susceptibility to BSE in transgenic mice expressing human prion protein. Nature 378:779–783

Collinge J, Sidle KCL, Meads J, Ironside J, Hill AF (1996) Molecular analysis of prion strain variation and the aetiology of 'new variant' CJD. Nature 383:685–690

DEFRA; Bovine Spongiform Encephalopathy in the UK: A Progress Report (2002) DEFRA Publications, London

Fraser H (1979) The pathogenesis of pathology of scrapie. in: Tyrell DAJ, (Eds.), Aspects of slow and persistent virus infections. Martinus Nijhoff, The Hague, 30–58

Gajdusek DC, Zigas V (1957) Degenerative disease of the central nervous system in New Guinea: the endemic occurrence of 'Kuru' in the native population. N Engl J Med 257:974–978

Gambetti P, Parchi P, Petersen RB, Chen SG, Lugaresi E (1995) Fatal familial insomnia and familial Creutzfeldt-Jakob disease: clinical, pathological and molecular features. Brain Pathol 5:43–51

Giese A, Kretzschmar HA (2001) Prion-induced neuronal damage. Curr Top Microbiol Immunol 253:203–217

Green AJ, Thompson EJ, Stewart GE, Zeidler M, McKenzie JM, MacLeod MA, Ironside JW, Will RG, Knight RS (2001) Use of 14-3-3 and other brain-specific proteins in CSF in the diagnosis of variant Creutzfeldt-Jakob disease. J Neurol Neurosurg Psychiatry 70:744–748

Hainfellner JA, Budka H (1999) Disease associated prion protein may deposit in the peripheral nervous system in human transmissible spongiform encephalopathies. Acta Neuropathol 98:458–460

Hainfellner JA, Parchi P, Kitamoto T, Jarius C, Gambetti P, Budka H (1999) A novel phenotype in familial Creutzfeldt-Jakob disease: Prion protein gene E200 K mutation coupled with valine at codon 129 and type 2 protease-resistant prion protein. Ann Neurol 45:812–816

Head MW, Tissingh G, Uitdehaag BMJ, Barkhof F, Bunn TJB, Ironside JW, Kamphorst W, Scheltens P (2001) Sporadic Creutzfeldt-Jakob disease in a young Dutch valine homozygote: atypical molecular phenotype. Ann Neurol 50:258–261

Head MW, Northcott V, Rennison K, Ritchie D, McCardle L, Bunn TJR, McLennan NF, Ironside JW, Tullo AB, Bonshek RE (2003) Prion protein accumulation in eyes of patients with sporadic and variant Creutzfeldt-Jakob disease. Invest Ophthalmol Vis Sci 44:342–346

Hill AF, Butterworth RJ, Joiner S, Jackson G, Rossor M, Thomas DJ, Frosh A, Tolley N, Bell JB, Spencer M, King A, Al-Sarraj S, Ironside JW, Lantos PL, Collinge J (1999) Investigation of variant Creutzfeldt-Jakob disease and other human prion diseases with tonsil biopsy samples. Lancet 353:183–189

Hill AF, Desbruslais M, Joiner S, Sidle KCL, Gowland I, Collinge J, Doey LJ, Lantos P (1997) The same prion strain causes vCJD and BSE. Nature 389:448–450

Hilton DA, Fathers E, Edwards P, Ironside JW, Zajicek J (1999) Prion immunoreactivity in appendix before clinical onset of variant Creutzfeldt-Jakob disease. Lancet 352:703–704

Hsiao K, Baker HF, Crow TJ, Poulter M,Owen F, Terwilliger JD, Westaway D, Ott J, Prusiner SB (1989) Linkage of a prion protein missense variant to Gerstmann-Straussler syndrome. Nature 338:342–345

Hunter N, Foster J, Chong A, McCutcheon S, Parnham D, Eaton S, MacKenzie C, Houston F (2002) Transmission of prion diseases by blood transfusion. J Gen Virol 83:2897–2905

Ironside JW (1998) Prion diseases in man. J Pathol 186:227–234

Ironside JW, Sutherland K, Bell JE, McCardle L, Barrie C, Estebeiro K, Zeidler M, Will RG (1996) A new variant of Creutzfeldt-Jakob disease: neuropathological and clinical features. Cold Spring Harbor Symp Quant Biol LXI:523–530

Ironside JW, Head MW, Bell JE, McCardle L, Will RG (2000a) Laboratory diagnosis of variant Creutzfeldt-Jakob disease. Histopathology 37:1–9

Ironside JW, Hilton DA, Ghani A, Johnston NJ, Conyers L, McCardle LM, Best D (2000b) Retrospective study of prion-protein accumulation in tonsil and appendix tissue. Lancet 355:1693–1694

Lasmezas CI, Fournier JG, Nouvel V, Boe H, Marce D, Lamoury F, Kopp N, Hauw JJ, Ironside J, Bruce M, Dormont D, Deslys JP (2001) Adaptation of the bovine spongiform encephalopathy agent to primates and comparison with Creutzfeldt-Jakob disease: implications for human health. Proc Natl Acad Sci USA 98:4142–4147

Lasmézas CI, Ironside JW, Chiach C, Demaimay R, Adjou KT, Hauw JJ, Dormont D (1996) BSE transmission to macaques. Nature 381:743–744

Lorains JW, Henry C, Agbamu DA, Rossi M, Bishop M, Will RG, Ironside JW (2001) Variant Creutzfeldt- Jakob disease in an elderly patient. Lancet 357:1339–1340

Mastrianni JA, Nixon R, Layzer R, Telling G, Han D, DeArmond SJ, Prusiner SB (1999) Prion protein conformation in a patient with sporadic fatal insomnia. New Engl J Med 340:1630–1638

McLean CA, Ironside JW, Masters CL (1997) Comparative neuropathology in kuru and new variant CJD. Brain Pathol 7:1247–1248

Pan T, Colucci M, Wong BS, Ruliang L, Liu T, Petersen RB, Chen S, Gambetti P, Sy M-S (2001) Novel differences between two human prion strains revealed by two-dimensional gel electrophoresis. J Biol Chem 276:37284–37288

Parchi P, Capellari S, Chen SG, Petersen RB, Gambetti P, Kopp N, Brown P, Kitamoto T, Tateishi J, Giese A, Kretzschmar (1997) Typing prion isoforms. Nature 386:232–233

Parchi P, Castellani R, Capellari S, Ghetti B, Young K, Chen SG, Farlow M, Dickson DW, Sima AAF, Trojanowski JQ, Petersen RB, Gambetti P (1996) Molecular basis of phenotypic variability in sporadic Creutzfeldt-Jakob disease. Ann Neurol 39:767–778

Parchi P, Chen SG, Brown P, Zou W, Capellari S, Budka H, Hainfellner J, Reyes PF, Golden GT, Hauw JJ, Carlton Gajdusek D, Gambetti P (1998b) Different patterns of truncated prion protein fragments correlate with distinct phenotypes in P102L Gerstmann-Straussler-Scheinker disease. Proc Natl Acad Sci USA 95:8322–8327

Parchi P, Gambetti P, Piccardo P, Ghetti B (1998c) Human prion diseases. in: (Kirkham N, Lemoine NR, Eds) Progress in Pathology, Edinburgh, Churchill Livingstone 39–78

Parchi P, Petersen RB, Chen SG, Autillio-Gambetti L, Capellari S, Monari L, Cortelli P, Montagna P, Lugaresi E, Gambetti P (1998a) Molecular pathology of fatal familial insomnia. Brain Pathol 8:539–548

Parchi P, Zou W, Wang W, Brown P, Capellari S, Ghetti B, Kopp N, Schulz Schaefer WJ, Krezscmar HA, Head MW, Ironside JW, Gambetti P, Chen SC (2000) Genetic influence on the structural variation of the abnormal prion protein. Proc Natl Acad Sci USA 97:10168–10172

Parchi P, Castellani R, Cortelli P, Montagna P, Chen SG, Petesen RB, Manetto V, Vnecak-Jones CL, McLean MJ, Sheller JR, Lugaresi E, Gambetti L, Gambetti P (1995) Regional distribution of protease-resistant prion protein in fatal familial insomnia. Ann Neurol 38:21–29

Parchi P, Giese A, Capellari S, Brown P, Schulz-Schaeffer WJ, Windle O, Zerr I, Budka H, Kopp N, Piccardo P, Poser S, Rojiani A, Streichenebrger N, Julien J, Vital C, Ghetti B, Gambetti P, Kretzschmar HA (1999) Classification of sporadic Creutzfeldt-Jakob disease based on molecular and phenotypic analysis of 300 subjects. Ann Neurol 46:224–233

Prusiner SB (1998) Prions. Proc Natl Acad Sci USA 95:13363–13383

Prusiner SB (1993) Genetic and infectious prion diseases. Arch Neurol 50:1129–1153

Scott MR, Will R, Ironside J, Nguyen HO, Tremblay P, DeArmond SJ, Prusiner SB (1999) Compelling transgenic evidence for transmission of bovine spongiform encephalopathy prions to humans. Proc Nat Acad Sci USA 96:15137–15142

Shimizu S, Hoshi K, Muramoto T, Homma M, Ironside JW, Kuzuhara S, Sato T, Yamamoto T, Kitamoto K (1999) Creutzfeldt-Jakob disease with florid-type plaques after cadaveric dura mater grafting. Arch Neurol 56:357–363

Somerville RA (1999) Host and transmissible spongiform encephalopathy agent strain control glycosylation of PrP. J Gen Virol 80:1865–1872

Spence MD, Knight RSG, Will RG (2002) First one hundred cases of variant Creutzfeldt-Jakob disease: retrospective case note review of early psychiatric and neurological features. Brit Med J 324:1479–1482

Telling GC, Parchi P, DeArmond SJ, Cortelli P, Montagna P, Gabizon R, Mastrianni J, Lugaresi E, Gambetti P, Prusiner SB (1996) Evidence for the conformation of the pathologic isoform of the prion protein enciphering and propagating prion diversity. Science 274:2079–2082

The BSE Inquiry (2000) Vol 1: Findings and Conclusions. London, The Stationery Office

Turner M (2001) Variant Creutzfeldt-Jakob disease and blood transfusion. Curr Opin Haematol 8:372–379

Wadsworth JD, Joiner S, Hill AF, Campbell TA, Desbruslais M, Luthert PJ, Collinge J (2001) Tissue distribution of protease resistant prion protein in variant Creutzfeldt-Jakob disease using a highly sensitive immunoblotting assay. Lancet 358:171–180

Wadsworth JDF, Hill AF, Joiner S, Jackson GS, Clarke AR, Collinge J (1999a) Strain-specific prion-protein conformation determined by metal ions. Nature Cell Biol 1:55–59

Wadsworth JDF, Jackson GS, Hill AF, Collinge J (1999b) Molecular biology of prion propagation. Curr Opin Genetics Develop 9:338–345

Wells GA, Scott AC, Johnson CT, Gunning RF, Hancock RD, Jeffrey M, Dawson M, Bradley R (1987) A novel progressive spongiform encephalopathy in cattle. Vet Rec 121:419–20

Wilesmith JW, Wells GA, Cranwell MP, Ryan JB (1988) Bovine spongiform encephalopathy: epidemiological studies. Vet Rec 123:638–44

Will RG, Ironside JW, Zeidler M, Cousens S, Estebeiro K, Alperovitch A, Poser S, Pocchiari M, Hofmann B, Smith PG (1996) A new variant of Creutzfeldt-Jakob disease in the UK. Lancet 347:921–925

Will RG, Alpers MP, Dormont D, Schonberger LB, Tateishi J (1999) Infectious and sporadic prion diseases. Prion Biology and Diseases ch12:465–507

Will RG, Matthews WB (1984) A retrospective study of Creutzfeldt-Jakob disease in England and Wales 1970–79 I: Clinical features. J Neurol Neurosurg Psych 47:134–140

Will RG, Zeidler M, Stewart GE, Macleod MA, Ironside JW, Cousens SN, Mackenzie J, Estibeiro K, Green AJE, Knight RSG (2000) Diagnosis of new variant Creutzfeldt-Jakob disease. Ann Neurol 47:575–582

Williams ES, Young S (1993) Neuropathology of chronic wasting disease of mule deer (odocoileus hemionus) and elk (Cervus elaphus nelsoni). Vet Pathol 30:36–45

World Health Organization (1996) Report of a WHO consultation on clinical and neuropathological characteristics of the new variant of CJD and other human and animal transmissible spongiform encephalopathies. World Health Organization, Geneva

Zanusso G, Farinazzo A, Fiorini M, Gelati M, Castagna A, Righetti PG, Rizzuto N, Monaco S (2001) pH-dependent prion protein conformation in classical Creutzfeldt-Jakob disease. J Biol Chem 276:40377–40380

Zanusso G, Righetti PG, Ferrari S, Terrin L, Farinazzo A, Cardone F, Pocchiari M, Rizzuto N, Monaco S (2002) Two-dimensional mapping of three phenotype associated isoforms of the prion protein in sporadic Creutzfeldt-Jakob disease. Electrophoresis 23:347–355

Zeidler M, Sellar RJ, Collie DA, Knight R, Stewart G, Macleod M-A, Ironside JW, Cousens S, Colchester ACF, Hadley DM, Will RG (2000) The pulvinar sign on magnetic resonance imaging in variant Creutzfeldt-Jakob disease. Lancet 355:1412–1418

The Epidemiology of Variant Creutzfeldt–Jakob Disease

P. G. Smith[1] · S. N. Cousens[1] · J. N. Huillard d'Aignaux[1] · H. J. T. Ward[2] · R. G. Will[2]

[1] Department of Infectious and Tropical Diseases, London School of Hygiene and Tropical Medicine, Keppel Street, London, WC1E 7HT, UK
E-mail: p.smith@lshtm.ac.uk

[2] National CJD Surveillance Unit, Western General Hospital, Crewe Road, Edinburgh, EH4 2XU, UK

Abstract Variant Creutzfeldt–Jakob disease (vCJD) was identified as a new disease in 1996. It was linked to infection with the bovine spongiform encephalopathy (BSE) agent although the epidemiological evidence for this was not strong, but later strain typing studies confirmed the association. The disease has affected predominantly young adults whose dietary and other characteristics are unexceptional compared to control groups, other than that all patients to date have been methoinine homozygous at codon 129 of the prion protein gene and the incidence has been about two times higher in the North of the UK. The number of cases in the 7 years after first identification of the disease has been considerably lower than initially feared, given the likely widespread exposure of the UK population to the BSE agent through contaminated beef products. Predictions of the possible future course of the epidemic have many associated uncertainties, but current mathematical models suggest that more than a few thousand cases is unlikely. Such modelling is limited by the absence of a test for infection with the vCJD agent. The development of a test that could be used on easily accessible tissue to detect infection early in the incubation period would not only advance understanding of the epidemiology of infection with the agent but would also aid the implementation of control measures to prevent potential iatrogenic spread.

1 Introduction

In this chapter we review first the events leading up to the recognition of variant Creutzfeldt-Jakob disease (vCJD) as a new human disease in 1996. The evidence linking vCJD to bovine spongiform encephalopathy (BSE) in 1996 is summarized, as are the studies that were conducted subsequently that established beyond reasonable doubt that there is a causal association between the two conditions, both being due to infectious agents that are currently indistinguishable. The characteristics of the cases of vCJD ascertained to date are discussed in relation to the classical epidemiological features of person, place and time and we review attempts to predict the future course of the vCJD epidemic. Finally, we discuss some of the outstanding questions that may be addressed by epidemiological studies and some of the challenges these will pose.

2 The Emergence of vCJD

2.1 The Identification of vCJD as a New Disease

The likelihood of the BSE agent being a human pathogen was assessed to be 'remote' by the scientific working party set up to advise the UK government soon after the recognition of the cattle epidemic in the UK (Working Party on Bovine Spongiform Encephalopathy 1989). This judgement was influenced by the prevailing view that the BSE epidemic was likely to have originated from a cattle-adapted form of the agent responsible for the transmissible spongiform encephalopathy (TSE) in sheep—scrapie. Scrapie was known to have been endemic in the UK and other sheep populations since the early eighteenth century (Parry 1983) and no evidence had accumulated that the consumption of tissues from scrapie-infected sheep had caused any disease in humans. In particular, the human TSE Creutzfeldt–Jakob disease (CJD) occurred at a similar frequency in the UK as in Australia and New Zealand, where the sheep populations had been kept free of scrapie. Also, case–control investigations have not shown any convincing association between the risk of CJD and the consumption of brains, the tissue likely to have the highest titres of the infectious agent in scrapie-affected animals (van Duijn et al. 1998; Wientjens et al. 1996).

However, because of the possibility that the BSE agent might be a human pathogen, precautionary measures were introduced to restrict potential human exposure to BSE. Cattle affected by BSE were banned from the food chain, as were some tissues from all cattle, so-called 'specified bovine offals', including brain and spinal cord. These tissues were thought likely, in infected animals, to have the highest titres of the infectious agent, based on what was known about scrapie, a disease for which there was a much longer history of experimental studies. In addition to the controls on human exposure to bovine tissues potentially infected by BSE, it was recommended that systematic national surveillance for CJD should be introduced in the UK as it was thought that if BSE did prove to be a human pathogen, the disease or diseases it would cause would be most likely to resemble CJD (Working Party on Bovine Spongiform Encephalopathy 1989). Thus, in May 1990, the National CJD Surveillance Unit was set up, with the remit to gather data on all human TSEs in the UK in order to detect any changes in the epidemiology of these diseases that might be compatible with the BSE agent having caused disease in humans. No specific indication was given as to the form that any such changes might be expected to take, other than of a possible overall rise in the incidence of human TSEs.

Interest in the epidemiology of CJD in the UK had preceded the onset of the BSE epidemic. Specific research investigations had previously been undertaken to ascertain the incidence of the disease in England

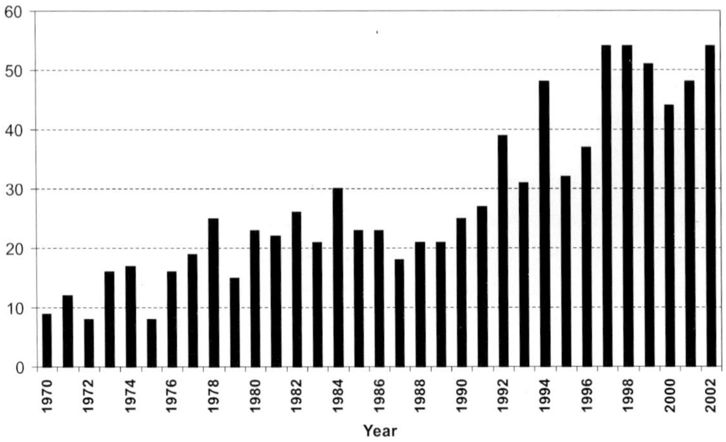

Fig. 1 Annual numbers of deaths from sporadic CJD reported in England and Wales, 1970–2002

and Wales, retrospectively for the period 1970–1979 (Will et al. 1986) and prospectively in the UK from 1980 to 1984 (Harris-Jones et al. 1988). Because neurologists see most cases of CJD and the number of consultants in this speciality in the UK was small—about 250—it was relatively easy to obtain information on all diagnosed cases, both retrospectively and prospectively, by direct liaison with these specialists. In this way it has been possible to examine trends in the numbers of cases of CJD in England and Wales since 1970 and in the UK since 1985. Fig. 1 shows the number of cases of sporadic CJD (excluding iatrogenic cases, familial cases and also cases of vCJD) reported annually in England and Wales since 1970.

The average annual number of cases identified increased from 10–15 a year in the 1970s, to 40–50 a year in the 1990s. Although this increase does not closely mirror the onset and temporal pattern of the BSE epidemic in cattle, an increase in the incidence of CJD was one of the principal epidemiological changes that might have been expected if BSE was a human pathogen. Closer study of the temporal trend in the incidence of sporadic CJD indicated that although there were increases in incidence in all age groups over the period 1970–2002, the most striking increases were in the older age groups, especially in those over the age of 60 years. This may be seen in Fig. 2 which shows age- and sex-specific incidence rates for the periods 1970–1989 and 1996–2002.

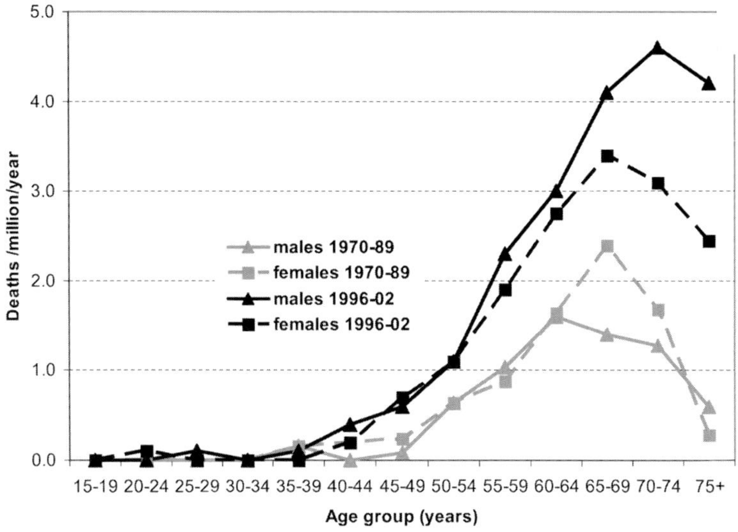

Fig. 2 Age- and sex-specific mortality rates from sporadic CJD (**a**) 1970–1989 (**b**) 1996–2002. (Data for 1970–1984 based on England and Wales only, thereafter data for all UK. Data from National CJD Surveillance Unit 2002)

Whereas in the earlier period the age-specific incidences fell substantially after the age of about 65 years, this decline was much less marked in the later period. Though the possibility that the overall increase in incidence was attributable to BSE could not be excluded, the favoured explanation has been that of improved ascertainment of cases. Similar increases in incidence have been reported in other countries, including in countries in which BSE has not been identified in cattle. Also, most of the increase was in older age groups in which other causes of dementia were much more common. These observations strongly suggested that much if not all of the increase in incidence might have been due to more complete ascertainment of cases in later periods rather than due to a true change in disease incidence. Such an effect might be expected given the much greater interest in CJD generated by the onset of the BSE epidemic and knowledge of the newly instituted national surveillance for CJD cases.

In parallel with the institution of a national surveillance system for CJD, an ongoing case-control investigation was set up in the UK with the aim of identifying risk factors for CJD and monitoring changes in the epidemiology of CJD that might be related to exposure to the BSE agent. As by the time cases are identified they are generally too sick to be questioned, the relatives of cases of CJD and those of 'controls' without CJD are interviewed using a structured questionnaire. The information collected includes full residence histories and data on a wide variety of other factors, including dietary habits, education, occupation and previous medical history. A finding of possible significance arising from these investigations was the observation in 1995 that cattle farmers seemed to be over-represented among the cases and this raised the suspicion that this might be related to their close proximity to cattle, some of which might be infected with BSE. Statistical calculations, based on the size of the farming population, indicated that the excess of cases (6 cases compared to 2.4 expected, P=0.03; 4 of these cases were on farms on which BSE had been reported compared to 0.58 expected, P=0.003) was unlikely to be due to chance and there were fears that this was the first sign that BSE had affected the human population (Cousens et al. 1997b; Gore 1995). However, parallel studies in other European countries with much lower incidences of BSE recorded a similar excess of CJD among farmers, which considerably weakened the evidence that the excess might be attributable to BSE (Cousens et al. 1997b). Subsequently, strain typing studies (see below) using brain material collected from two of the UK cases in farmers indicated that the cases were indistinguishable from sporadic CJD (Bruce et al. 1997).

In 1995, at about the same time as the observation of cases of CJD in farmers was reported and of much greater significance, two teenage cases of CJD were identified in the UK (Britton et al. 1995; Bateman et al. 1995). Although sporadic CJD had been reported in young persons previously, such cases were very rare and two cases in the UK in the same year raised serious concerns that this might be the first sign that the epidemiology of the disease was changing. It was difficult, however, to draw firm conclusions on the basis of just two cases. Later in 1995 and in the early months of 1996 further cases of CJD in young persons were notified to the National CJD Surveillance Unit and, by March 1996, it was clear that a highly significant change had occurred with respect to the epidemiology of CJD in young people. Table 1 shows the numbers of cases of CJD reported between 1970 and March 1996 in persons under the age of 45 years.

For the 15-month period leading up to March 1996 there were six cases of CJD reported in persons under the age of 30 years, compared to

Table 1 Cases of CJD in the UK; patients dying, aged less than 45 years between 1970 and March 1996 (excluding known iatrogenic and inherited cases)

	Age at death (years)			
	<30	30–34	35–39	40–44
1970–1979	0	2	3	2
1980–1984	1	1	3	1
1985–1989	0	0	3	3
1990–1994	0	0	1	2
1995–1996 (March)	5 (1)[a]	2 (1)	0	1

[a] The number of patients alive is given in parentheses.

only one such case in the preceding 25 years (excluding iatrogenic and familial cases). This in itself was highly statistically significant, but the importance of the finding was further enhanced by the recognition of a distinct neuropathological pattern in these young cases. This pattern was consistent between cases and was clearly distinguishable from that seen in all the other cases of sporadic CJD examined since 1990 as part of surveillance activities (Will et al. 1996). These observations led the Spongiform Encephalopathy Advisory Committee, which had been set up to advise the UK Government on scientific aspects of BSE and other TSEs, to conclude that the most probable explanation for this 'new variant' of CJD in young people was that they were due to human exposure to the BSE agent.

2.2
Evidence Linking vCJD to BSE in March 1996

The announcement by the UK Government of the likely link between BSE and cases of an apparently new disease, vCJD, had a profound and immediate effect. It was closely followed by a worldwide ban on British beef products and a ban on the consumption in the UK of cattle over the age of 30 months. At the time of the announcement, however, the evidence supporting the hypothesis that vCJD was due to BSE was not strong epidemiologically and rested to a considerable extent on the biological plausibility of a causal association.

The epidemiological evidence for the association was based on two observations. Firstly, vCJD was confined to the UK; at least as far as could be ascertained at that time, from whence the overwhelming majority of cases of BSE had been reported. Secondly, the cases of vCJD then reported had their disease onset in the mid 1990s, about 5–10 years after the peak of the BSE epidemic in cattle in the late 1980s and early 1990s. Incubation periods of this order had been observed with respect to iatrogenically-transmitted cases of CJD (following administration of contaminated batches of human growth hormone) and it was plausible, therefore, that BSE-induced human disease might have a similarly long incubation period. These observations, either alone or together, cannot be regarded as other than as being consistent with a causative association and certainly did not constitute 'proof beyond reasonable doubt'. Stronger epidemiological evidence might have come from the ongoing case–control investigation, which seeks evidence of differences in the past habits and practices of cases of CJD and controls without CJD. When the responses for vCJD cases and their controls were analysed, there were no striking differences between them, for example, with respect to the consumption of the types or quantities of beef products. However, for nearly all cases and controls there was a history of eating beef. It is likely that many different beef products were contaminated with the BSE agent and the finding of no differences between cases and controls in this respect could not be taken as strong evidence against a causal link between vCJD and BSE. Inaccuracies in recall are likely to have occurred with respect to information on diet 10 or more years previously, especially because this information was sought from close relatives rather than directly from the cases.

A possibility that could not be completely excluded in 1996 was that the 'new' disease had been identified as a direct consequent of the surveillance activities in the UK since 1990. The disease might have been present for much longer, including outside the UK and before the BSE epidemic, but without being diagnosed. However, it is hard to explain, on this basis, why it was not until 1995 that the first cases were diagnosed, following the introduction of enhanced UK surveillance in 1990. Furthermore, enhanced CJD surveillance had been introduced in a number of other European countries in 1993, none of which had identified any cases of vCJD by March 1996.

2.3
Subsequent Evidence Supporting the Causative Association

The evidence that has accumulated since 1996 on the link between BSE and vCJD has strongly supported the notion that the association is causal. Direct evidence of the route of transmission from cattle to humans is lacking but dietary exposure to contaminated beef products seemed the most plausible hypothesis.

The total number of vCJD cases worldwide is currently less than 150 and no cases have been reported with onset before 1994. Ten cases have now been reported outside of the UK (France, six; Ireland, one; Italy, one; USA, one; Canada, one). However, these have either been in individuals who have spent considerable amounts of time in the UK during the relevant period for exposure or who may have been exposed to BSE-contaminated products locally. For example, during the period 1980–1995 about 10% of British beef was exported to France (Alperovitch and Will 2002) and although few of the French vCJD cases report spending time in the UK, the possibility of exposure to BSE-contaminated products in France cannot be excluded; indeed, it seems likely (Alperovitch and Will 2002).

The strongest evidence for a causative link between BSE and vCJD comes not from epidemiological studies but from strain typing studies (reviewed in the chapters by Calavas et al. and Silveira et al. in this volume). Experimental studies using a mouse strain typing system (Bruce et al. 1997) and more recently developed molecular biochemical methods (Hill et al. 1997) were both unable to distinguish any differences between the characteristics of the infectious agents from brain material taken from cases of BSE and of vCJD. That the two causative agents are indistinguishable by these techniques has been taken to establish, beyond reasonable doubt, that vCJD is 'human BSE' (Almond and Pattison 1997).

3
Epidemiological Studies of vCJD

The epidemiological approach to the study of a disease generally starts by focussing on three fundamental characteristics—summarized as 'person, place and time'. Knowledge about the aetiology of disease and possible transmission mechanisms for an infectious disease can often be derived from studies of differences between the characteristics (such as occupation, diet, and exposure to different environments/agents) of those with and without the disease under study. Furthermore, study of temporal and spatial trends and patterns of disease may also be highly informative. So far, the number of cases of vCJD has been relatively small and this limits the potential for epidemiological investigation and inference. Nonetheless, some striking features have been identified that must hold important clues to disease aetiology. We summarize these epidemiological investigations and findings in the sections that follow.

3.1 Person

3.1.1 Age-Related Susceptibility/Exposure

One of the most striking features of vCJD, and indeed the characteristic that contributed most to the early recognition of the disease is the age distribution of the cases. The age-specific incidence rates for males and females are shown in Fig. 3.

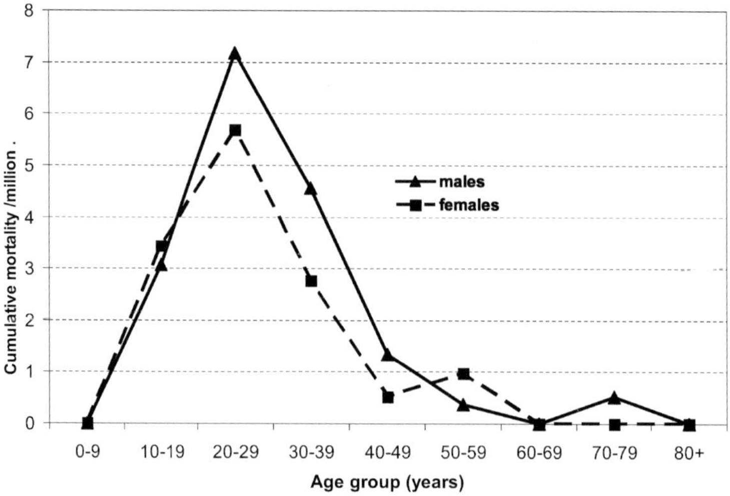

Fig. 3 Cumulative age- and sex-specific mortality rates for vCJD in the UK up to 31 December 2002

In both sexes, the highest death rates have been in the age group 20–29 years. The youngest case had onset at age 12 years and at the oldest at age 74 years. We have commented above on the possible under-ascertainment of cases of sporadic CJD in old people. There are concerns that cases of vCJD may also be missed in old people, especially in age groups in which other causes of dementia are common and post mortems (and especially neuropathological investigations) are relatively uncommon (Lowe 2000). However, it seems unlikely that individuals with dementia at younger ages (e.g., 35–55 years) would not be subject to more detailed investigation. Thus, the lower incidence in persons aged 35–55 years compared to those aged 20–29 years is probably a real feature of the epidemiology of the disease.

There have been concerns that, because children may not be referred to the neurologists who would be expected to see and report cases of vCJD in adults, cases of vCJD in children may have been missed. To address this possibility, a special surveillance system was set up among paediatricians in the UK to detect and investigate all cases of progressive intellectual and neurological deterioration in children. The initial report for that surveillance programme identified only a small number of older children with vCJD all of whom had been notified to the National CJD Surveillance Unit. This finding suggested that the deficit of cases in children was unlikely to be due to poor ascertainment in that age group (Verity et al. 2000).

Thus, in summary, it seems unlikely that the apparent variation in the risk of vCJD at different ages is wholly due to biases in the ascertainment system, though the dearth of cases in the older age groups may be partially attributable to such a bias. Another possibility is that the incubation period of vCJD is age dependent, older persons having substantially longer incubation periods. Evidence of such an effect has been sought for cases of kuru. Estimated average incubation periods were not clearly related to age and such differences as were seen, which differed in direction for men and women, were thought more likely to be related to effects of exposure–dose rather than age as such (Huillard d'Aignaux et al. 2002). Overall, therefore, the age distribution of cases of vCJD probably reflects either variations in age-related susceptibility to disease or variations in age-related exposure to the causative agent or, perhaps most likely, some combination of these two possible effects. Unfortunately, relatively little experimental work has been done to examine age-related variations in susceptibility to experimental TSE infections in animal models. There is some evidence of age-related variation in incubation periods in experimental mouse models but the effects described are not very large and because of the great differences in life span it is difficult to assess the relevance of this work for humans (McLean and Bostock 2000).

Mathematical modelling of the BSE epidemic has suggested that cattle in the first 2 years of life must have been more susceptible to BSE infection from contaminated meat and bone meal than were older animals (Anderson et al. 1996). Furthermore, it has been suggested that the reason that the BSE epidemic originated in the UK, rather than in any of a number of other countries with similar animal husbandry and cattle feeding practices, was that the UK was the first country to start feeding meat and bone meal to very young calves and there may have been greater susceptibility to BSE in this age group (Horn et al. 2001). However, direct experimental observations on age-related susceptibility in cattle are lacking. Studies are ongoing to assess age-related susceptibility to experimentally induced BSE in sheep but no results are yet available.

That many of the cases of vCJD have arisen in young adults led to the suggestion that there might be an association with vaccination in childhood and the age at onset largely reflected the length of the incubation period. However, a detailed investigation of the dates of birth of cases related to the usual ages at which vaccines are given did not lend credence to the hypothesis that vaccines were important in the aetiology of vCJD (Minor et al 2000).

The peak of the BSE epidemic in cattle corresponded with the time when some of the cases, but by no means all, were in their childhood and teens and this group might have increased susceptibility, but there is no direct evidence of this. Variations in age-related exposure to the BSE agent may be reflected in the age distribution of vCJD cases. Cheap meat products, such as minced beef and hamburgers, are probably most likely to have been most contaminated with the BSE agent. This is because of the likely greater incorporation in such products of mechanically recovered meat (MRM) which may have been contaminated with 'high risk' tissues such as spinal cord and dorsal root ganglia (Det Norske Veritas Consulting 2002). The consumption of such products was probably more common in children and teenagers than in older adults (Gore 1997). This may account, at least in part, for the higher incidence of vCJD in young adults but consumption of such products is not confined to young people and the steep fall in incidence in those aged 35–55 years is difficult to explain on this basis alone.

Some cheap meat products were consumed in greater quantities by males than females and for this reason it has been suggested that an excess of cases of vCJD might be expected in males (Cooper and Bird 2002). Of the cases in the UK to date, 53% have been in males (National CJD Surveillance Unit 2001), which is comparable with the excess predicted, but is also compatible with equal incidence in males and females.

3.1.2 Genetic Susceptibility

Polymorphisms of the prion protein gene have been associated with large variations in susceptibility to TSE diseases in experimental and natural history studies, most notably in sheep. Polymorphisms at codon 129 of the human PrP gene are also associated with variations in the occurrence of sporadic CJD. Whereas about 70% of cases of sporadic CJD are in persons who are methionine homozygous (MM) at codon 129 only about 40% of the general population are methionine homozygous at codon 129 (Table 2) (Alperovitch et al. 1999).

To date all cases of vCJD that have been genotyped have been methionine homozygous at codon 129. It may be that other genotypes are not susceptible to infection with the BSE agent, at least with respect to the dose of the agent to which the population has been exposed. A more pessimistic scenario is that those who are valine homozygous or who are heterozygous at codon 129 may be susceptible but have, on average, longer incubation periods than those who are MM. Observations on kuru (Lee et al. 2001) and on the epidemics of iatrogenic CJD following the administration of contaminated human growth hormone (Huillard d'Aignaux et al. 1999) indicate that the average incubation period may vary in subgroups of the population according to the genotype at codon 129. In most experimental TSE systems longer incubation periods are

Table 2 Genetic susceptibility to CJD according to polymorphisms of the PrP gene (Alperovitch et al. 1999)

	Distribution of genotypes		
PrP gene codon 129	General population	Sporadic CJD	vCJD
MM	39%	71%	100%
VV	11%	16%	0
MV	50%	13%	0

associated with lower attack rates. Thus, even if some of those who are valine homozygous or who are heterozygous at codon 129 do come down with vCJD, the proportions of the population affected in these subgroups may be smaller than that in those who are MM (see below).

It has also been suggested that other genetic polymorphisms, yet to be identified, and outside the PrP coding region many influence susceptibility to vCJD. Such effects on the length of incubation periods have been shown in some experimental mouse TSE studies (Lloyd et al. 2001). Recent publications relating susceptibility to vCJD to the HLA system have given conflicting results (Jackson et al. 2001; Pepys et al. 2003; Laplanche et al. 2003).

3.1.3 Case–Control Investigation

For many rare diseases the case–control approach has been a powerful epidemiological method to identify factors that affect the risk of disease. Interviews are conducted with cases of disease and with controls who do not have the disease of interest but who are similar to the cases with respect to characteristics such as age and sex. Cases and controls are asked questions about their past histories with respect to factors that may be related to the aetiology of the disease, and thus for which different responses would be expected from cases and controls. For vCJD it has been necessary to interview close relatives of cases and controls, as the cases are generally too sick to be interviewed by the time a diagnosis of vCJD is suspected. Such studies present many challenges and the avoidance of bias is a major consideration. Attempts to ascertain histories of diet and other factors relevant to possible exposure, 10 or more years in the past, and from relatives, are highly prone to inaccuracies. Furthermore, the selection of a suitable control group is often difficult. In the studies of sporadic CJD referred to above, other hospital patients without CJD were used for the selection of the control group. For some potential risk factors such a group provides a suitable control population. But, for example, when investigating histories of previous surgical procedures, the past histories of such procedures among hospitalized patients may be atypical of the general population and the use of such a group as controls may bias against finding relevant differences between case and controls. As iatrogenic factors are potentially important in the study of vCJD, attempts have been made to select controls from a group more representative of the general population by selecting at random persons from age and sex registers on general practitioner patient lists. However, this has proved logistically complex and has produced a very low response rate. A revised strategy has therefore been adopted to select a random sample of the general population using a database of residential addresses who are then visited by specially trained interviewers. In addition, the use of a control group consisting of persons nominated by the relatives of vCJD cases is also being explored. This work is ongoing and thus the case-control information assembled so far has to be interpreted with caution because of underlying potential biases.

Diet. Excesses of reported consumption of some meat products by cases have been found. In recent analyses, cases were reported to have eaten sausages, burgers, faggots and products containing MRM more frequently than community controls (National CJD Surveillance Unit 2002). However, care has to be taken in interpreting these findings, as there is considerable scope for recall bias with respect to dietary histories. The relatives of cases are generally well aware of hypotheses with respect to routes of exposure, as these have been much discussed in the popular media, and this may bias the responses they give. In order to examine the possibility of recall bias in relation to dietary history, cases of vCJD were compared with another group of controls, consisting of people referred to the National CJD Surveillance Unit with suspect vCJD who were subsequently determined to have an alternative diagnosis. Cases were reported to have eaten beef, burgers and sausages more frequently than these controls. However, these differences were relatively small and not statistically significant. Thus, the dietary data from the case-control investigation are very difficult to interpret.

Occupation. Data from the case-control study have been analysed to seek evidence that those with a history of particular occupations might be at increased risk of vCJD. In particular, evidence has been sought that cases might have worked more frequently in medical, dental or associated professions, in animal, pharmaceutical or research laboratories, in farming or veterinary medicine, in the meat or catering industry, or in other occupations involving animal products. In the analyses most recently reported, no evidence was found of associations with any of these occupations (National CJD Surveillance Unit 2002).

Medical and Surgical Risk Factors. A higher proportion of community controls were reported to have undergone some sort of operation/surgical procedure in the past compared with the cases, but this was not statistically significant. There was no evidence to suggest that particular operations/surgical procedures were associated with increased risk of vCJD. However, these analyses were based on relatively small numbers and the confidence intervals around the odds ratio estimates were wide. These findings therefore have to be interpreted with caution. Furthermore, if there has been iatrogenic transmission of vCJD and the incubation period is long, evidence of such transmission may not emerge until later in the epidemic as it would be expected that the cases of vCJD seen to date would be from the primary bovine to human transmission.

In summary, with the exception of the factors noted above, the case–control investigation has so far identified no substantial differences between reported 'exposures' for cases and controls across a wide variety of factors, including diet, occupational and medical histories (National CJD Surveillance Unit 2002).

3.2 Place

3.2.1 Regional Variations in Incidence

The geographical distribution of BSE in the UK has been heterogeneous. The epidemic has shown clustering with respect to the number of cases affecting particular herds, most herds having had relatively few cases but some having had a disproportionately large number of cases. Also, the distribution of BSE cases in different parts of the UK has shown marked variation with, overall, a substantial excess of cases in the south of England (Wilesmith et al. 1992). It is unclear, however, if and how these differences might be expected to correlate with geographical differences in the incidence of vCJD. Where an animal was reared probably bears little relation to where meat from that animal was eaten. Nonetheless, the geographical distribution of vCJD cases in the UK has shown some intriguing heterogeneity that may provide clues to disease risk factors.

Initial observations on the distribution of vCJD in different regions of the UK did not find evidence of significant heterogeneity. However, a trend of an excess of cases in the north of England and in Scotland be-

Table 3 Comparison of cumulative incidence of vCJD in the 'North' of the UK (excluding Northern Ireland) with that in the 'South'

Region	Population aged 10 years and above at the 1991 census	Number (rate/million) of vCJD cases by place of residence in 1991	
'North' (North West, Yorkshire and Humberside, Northern, Scotland)	16.6 million	53	(3.19)
'South' (South West, South East, Wales, West Midlands, East Midlands, East Anglia)	31.2 million	57	(1.83)
Total (rate ratio)[a]	47.8 million	110	(1.62)

[a] North versus South, adjusted for age and sex.

came apparent in more recent analyses (Cousens et al. 2001). An approximately 60% greater incidence in northern parts of the UK has persisted and is highly statistically significant (Table 3).

There are undoubtedly differences in diet between different parts of the UK and such data as exist to document these differences have been examined to seek possible explanations for the gradient in vCJD incidence. The results have been inconclusive and although an association was found with some types of products considered more likely to be potentially contaminated with the BSE agent, analyses of other data sets did not support this inference and the reason for the regional variation in incidence is currently unexplained. It seems most likely that it is related to dietary habits. Differences in genetic susceptibility remain a possible, but perhaps unlikely, explanation. Other analyses suggest it is not likely to be an artefact of ascertainment resulting from the location of the National CJD Surveillance Unit in Scotland (Cousens et al. 2001).

3.2.2 Local Clustering of Cases

Several analyses have been conducted to examine whether the geographical distribution of individual cases of vCJD is random with respect to specific potential risk factors or with respect to other cases of the disease. Following newspaper reports of an apparent cluster of cases in the vicinity of a rendering plant, a comprehensive analysis was conducted of the places of residence of all cases of vCJD in relation to all rendering plants operating around the time of the peak of the BSE epidemic. Circles of different radii were drawn around each plant and the number of cases of vCJD, whose place of residence in 1990 had been within such a circle, was compared to the size of the total population resident in the circle—obtained from census data. Circles of radii 1, 5, 10, 20 and 50 km were used, but in no instance was a significant excess of vCJD cases found, compared to the rate among those residing distant from a plant (Cousens et al. 1999).

Further analyses, examining the distribution of cases of vCJD in relation to each other, sought evidence of local clusters of disease. There was no strong a priori reason to suspect such clustering if dietary exposure to contaminated beef products was widespread throughout the UK. Overall, little evidence of such clustering was found, with one exception. A cluster of five cases was identified among patients who had lived within 5 km of each other in 1991—the time point chosen for one of the main analyses. This finding was statistically significant (Cousens et al. 2001) and prompted a local investigation of the possible reasons for the excess.

A search was made for possible common features or characteristics that might further link the five cases and that might give clues to the mode of transmission of vCJD in this community. A wide variety of possible common factors were investigated (Bryant et al. 2001). The one that seemed to link four of the five cases most strongly was the purchase of meat from two local butchers, who had used particular practices in preparing cattle carcases. After slaughter, it was their practice to first remove and split the head of the animal and to extract the brain, which was sold locally. The rest of the carcase was then cut up on the same bench and using the same instruments as been used on the head. Although none of the cases gave a history of purchasing or eating brain, it was hypothesised that carcase meat could have been contaminated with brain material during preparation. The investigators conducted a case–control investigation to assess the significance of the association of the cases with the particular butchers and obtained strongly positive results. Although this investigation has been the only one to date in which an apparent route of transmission was identified, a number of issues were left unresolved. First, it is likely that the highest titres of infectivity will have been in the brains of the slaughtered cattle, but neither these cases nor other cases have reported consumption of this material. Second, the investigation was, necessarily, of a small number of cases and it is unclear to what extent the findings might be generalized with respect to modes of transmission relevant to other cases. It seems unlikely that this particular butchery practice was confined to the geographical area of the cluster, though more widespread studies of butchery practices are ongoing, in which case other clusters might have been expected, but have not been found. Thirdly, if exposure was truly due to contamination of carcase meat, it suggests that the infecting dose is likely to have been quite small. This is difficult to reconcile with the relatively small size of the epidemic of vCJD to date and the very large number of infected cattle that must have gone into the human food chain (Anderson et al. 1996).

3.3 Time

The link between BSE and vCJD was postulated in 1996 following the recognition of the first 10 cases of vCJD (Will et al. 1996). It was known at that time that very large numbers of BSE-infected cattle had gone into the human food chain, including many thousands before specific controls were imposed to prohibit specified bovine offals, in particular brain and spinal cord, from the UK diet. Given the likelihood that a high proportion of the UK population had been exposed to contaminated beef products, there were fears that the consequent epidemic of vCJD may affect very large numbers, possibly hundreds of thousands, of people. Seven years have now passed since the first cases of vCJD were described and a total of less than 150 cases have been diagnosed to date in the UK. There are thus some grounds for optimism that the human toll may be substantially less than originally feared. However, there are still many uncertainties about the way in which the epidemic is likely to evolve and accurate predictions of future case numbers are impeded by the lack of a test for pre-clinical infection. Such predictions of the future course of the epidemic as have been made have been mostly based on mathematical models, though some have rejected the utility of this approach because of the many assumptions it is necessary to make in their construction (Lloyd et al. 2001). Mathematical models have been used to make both long- and short-term projections.

3.3.1 Short-Term Projections

In the early phases of epidemics of infectious diseases, the number of cases often tends to increase in an approximately exponential fashion, and the speed of growth is described by the percentage increase from one time period to the next. Farrington and colleagues analysed quarterly data on deaths and onsets from vCJD to measure the rate of increase over time, and devised a method which allowed for reporting delays between the onset of first clinical symptoms of vCJD and the diagnosis of the disease. In early analyses of both deaths and onsets the rate of increase was compatible with an increase of about 20% a year from 1995 onwards (Andrews et al. 2000). However, the most recent analyses have shown evidence of departures from a model of simple exponential increase. In these analyses a model with a quadratic component fits the data better than one with a simple (linear) exponential increase (Andrews et al. 2003). This may be interpreted to indicate that the rate of increase in the numbers of cases is slowing and may even have peaked. The number of deaths in each year since the start of the epidemic is shown in Table 4.

Table 4 Number of deaths from vCJD in the UK by year up to the end of October 2003

Year died	Number of deaths
1995	3
1996	10
1997	10
1998	18
1999	15
2000	28
2001	20
2002	17
2003 (to 1 September)	16
Total deaths	137[a]
Cases alive	6
Total cases	143

[a] Includes 36 without neuropathological confirmation, classified as probable cases on the basis of clinical signs and symptoms.

The year with most deaths was 2000 and the number has declined in each of the 2 following years. This is compatible with the epidemic having peaked around 2000, but it is too early to draw this conclusion with any confidence, as there are likely to be statistical fluctuations in the numbers from year to year. Nonetheless, it is an encouraging sign.

Of concern is the possibility that the epidemic curve may not be unimodal. As discussed above, no cases of vCJD have been discovered to date who are other than methionine homozygous at codon 129 of the PrP gene. If other genotypes are associated with longer incubation periods (as suggested by observations on kuru and growth hormone-related cases of CJD) then subsequent epidemics in these genotypes may yet appear.

3.3.2 Predicting the Total Size of the Epidemic

The method discussed in the section above does no more than analyse short-term trends in the occurrence of disease and may be useful for predictive purposes no more than a year or two in advance—and even then there are substantial uncertainties. For many reasons it would be desirable to have an assessment of the likely size of the whole epidemic. Not least among the reasons for this is to be able to plan for the burden the epidemic will place on the health care system and also to be able to plan measures to prevent secondary transmission that are proportionate to the likely magnitude of this risk. Considerably more information is required than we currently have in order to make predictions of this kind with any precision. However, several groups have constructed mathematical models of the epidemic, which make explicit the underlying assumptions in the predictive process. There is an emerging consensus among these groups that the vCJD epidemic is unlikely to be of the enormous magnitude feared when the disease was first described. The number of cases to date is compatible with the final size of the epidemic being of the order of a few hundred cases, and even more than a few thousand cases looks unlikely based on the current models (Huillard d'Aignaux et al. 2001; Valleron et al. 2001), although others have put the upper confidence interval higher (Ferguson et al. 2002), but in more recent work by the same group the estimates have been lowered substantially (Ghani et al. 2003). However, such models are limited by the assumptions on which they are based and must be interpreted with caution.

The statistical modelling methods used have been similar to those developed to predict the evolution of the AIDS epidemic, based on a method called back-calculation in which observations on the numbers of cases at different times are used to infer the numbers of infections that must have occurred historically (Gail and Brookmeyer 1988). These estimates of infections are then used to make forward predictions of cases of disease yet to occur. In order to predict the course of an epidemic of an infectious disease it is necessary to know how many people have been infected, when they were infected and what the distribution of incubation periods (time from infection to disease onset) is for the disease. For vCJD we can make a reasonable estimate of when infection of the human population occurred, assuming this relates to variations in the prevalence of BSE in the cattle population at different times. However, we are ignorant of the numbers of humans infected and of the incubation period distribution.

The only firm data that we have currently on the occurrence of vCJD is the number of cases that have been ascertained month-by-month since the start of the epidemic. The number of cases that have incidence at a particular point in time (say, a particular month) is composed of persons infected at different points in time whose clinical symptoms manifest at the particular point in time. We may express this mathematically as follows. The number of cases with onset at time t will consist of the number infected at time $t-1$ who have an incubation period of length 1, plus the number infected at time $t-2$ who have an incubation period of length 2, plus the number infected at time $t-3$ who have an incubation period of length 3, and so on.

The number of persons infected at different times will have depended upon a number of factors, including: the amount of infected bovine tissue entering the human food chain, how different doses of the infectious agent were distributed between individuals and what the form of the dose–response relationship was between exposure to infected material and the likelihood of infection being established. We have no information on most of these factors. But, although we do not know the numbers of persons infected at different times we can estimate the relative numbers of persons infected at different times. This may be expected to be proportional to estimates of the number of BSE cases occurring at different times or the number of infected animals, or amount of infected bovine tissue (volume×infectivity titre), going into the food chain.

Importantly, we also have little information about the distribution of incubation periods (the periods from infection to the onset of disease) in humans. For many infectious diseases the shapes of the incubation period distribution have been studied, which can be done with precision when the time of exposure to the infection is known. Typically, the incubation period distribution has a single peak (is unimodal) and is right-skewed, that is the incidence rises more rapidly to a peak following exposure than it declines after the peak (Sartwell 1950). Often, if the logarithm of the incubation periods is taken the resulting transformed distribution is close to a 'Normal' distribution—in this case the incubation period distribution is said to follow a log-normal distribution (Sartwell 1950). Other more complex mathematical distributions, with the same property of being unimodal may also be used to approximate the incubation period distribution. In the construction of mathematical models to predict the evolution of the epidemic it has been necessary to postulate a wide variety of different forms for the incubation period distribution and, in each case, to examine how different assumptions affect the predictions. In this way, ranges of predictions have been made, depending on the assumptions made.

Clearly, a critical assumption is the average length of the incubation period. If, for example, we assume this is of the order of 10 years, it would be reasonable to postulate that the number of cases of vCJD seen to date is a substantial proportion of all the cases due to bovine-to-human transmission. This is because the peak of the BSE epidemic in the UK was around 10 years ago. If the average incubation period is much longer than 10 years then the cases we have seen to date may be only a small proportion of all cases and the eventual epidemic size will be much larger. Unfortunately, for vCJD we do not know the incubation period distribution and, in particular do not know its average length or how much variation there is around that average. For cases of kuru and for cases of growth hormone-related CJD it has been estimated that the average incubation period must have been of the order of 10 years (Huillard d'Aignaux et al. 2002; d'Aignaux 1999), but with substantial variation around that with respect to individual cases. Cases of kuru have been described, for example, for which it is thought the incubation period is substantially in excess of the average (Prusiner et al. 1982) and up to 40 years or more (Liberski et al. 1997). For both of these diseases, however, the transmission of the infectious agent is thought to have been from human-to-human (through cannibalism for kuru and through iatrogenic transmission for growth hormone-related CJD). A common finding for TSEs is that the incubation period when crossing into a new species is longer than subsequent transmission within the same species. Thus, it is possible that kuru and growth hormone-related cases of CJD may be poor models when considering the likely behaviour of BSE when transmitted from cattle to humans and, in particular, it may be that the average incubation period is considerably in excess of 10 years. Thus, in the mathematical modelling, some have assumed that the average incubation period may be up to 25 years (Cousens et al. 1997a), while others have allowed for the possibility that the average may exceed the normal human lifespan (Huillard d'Aignaux et al 1991; Ferguson et al. 1999).

Estimates of the possible size of the epidemic, that were made relatively soon after vCJD was first described, ranged from less than 100 cases to over 100,000 cases (Cousens et al. 1997a; Ghani et al. 1998). As time has gone on and the increase in cases has not been as rapid as was initially feared predictions of the maximum size of the epidemic have diminished, as discussed above. Indeed there is some, albeit weak, evidence that the peak of the epidemic may have been reached (Table 4) (Andrews et al. 2003).

It is important, however, to stress the caveats associated with the mathematical modelling conducted to date. First, a fundamental assumption has been that the incubation period distribution is unimodal. The strongest grounds for questioning this assumption relate to the observation that all cases of vCJD who have been genotyped to date have been methionine homozygous at codon 129 of the PrP gene. Thus, the predictive models relate only to the approximately 40% of the population who are of this genotype. If the other 60% of the population are not completely resistant to infection with the vCJD agent, but disease in these individuals is associated with longer average incubation periods, then epidemics of cases in these groups may yet occur. Observations of kuru and growth hormone-related CJD indicate that the incubation period in these diseases is influenced by the PrP genotype at codon 129. If, however, longer incubation periods are associated with greater resistance to infection then it might be speculated that even if disease does occur in those other than methionine homozygous (MM) at codon 129, the size of these epidemics may total less than one-and-a half times the epidemic in those who are MM (as those who are MM comprise about

40% of the population and those who are not comprise 60%). If genes other than *PRNP* are associated with susceptibility to, or affect the incubation period of, vCJD, as has been reported to be the case in studies of TSEs in mice (Lloyd et al. 2001), then it is possible that even within those MM at codon 129, there may be subgroups of different susceptibility and/or different average incubation periods. If this were so, the cases seen so far would tend to be from the most susceptible subgroups. If there are many other genes controlling susceptibility, then the assumption of an overall unimodal distribution is probably reasonable; however, if only a few other genes are involved the overall distribution might be multi-modal (and would invalidate the prediction methods that have been used).

A second assumption in the modelling is that the ascertainment of cases is near complete. While this is likely to be true for cases of vCJD in young adults, there remains the possibility that cases in old people are being missed because of the small proportion of those dying with dementia who are subject to post-mortem including a neuropathological examination. Furthermore, while the spectrum of disease in those with vCJD is reasonably well defined, it is possible that human infection with the BSE agent may result in other forms of disease not yet identified as associated with BSE (for example, in those other than MM at codon 129 of the PRNP gene). Recent experimental studies with transgenic mice have raised the possibility that some cases of sporadic CJD may be BSE associated (Asante et al. 2002) and this was one of several postulates for a recent rise of cases of sporadic CJD in Switzerland (Glatzel et al. 2002). However, at present these associations are speculative rather than proven.

Thirdly, most of the modelling has assumed that exposure of the human population has been from bovine products contaminated with BSE. Although definite evidence is lacking, there is a theoretical possibility that sheep were infected with BSE (Kao et al. 2002). They have been shown to be susceptible to infection experimentally and are known to have been fed the same meat and bone meal that caused the BSE epidemic in cattle, albeit in considerable smaller quantities. Furthermore, if BSE behaves like scrapie in sheep, the agent may have persisted in the sheep population through horizontal transmission. The feed controls would therefore have been less effective in sheep than they have been in cattle, in which horizontal transmission has not been demonstrated to occur. If the possibility of infection through contaminated ovine meat is taken into account, the predicted possible size of the vCJD epidemic increases (Ferguson et al. 2002).

Fourth, most of the modelling to date has considered only bovine-to-human transmission. CJD is known to have been transmitted iatrogenically (Brown et al. 2000) and there are considerable grounds for concern regarding the possibility of such transmission of vCJD, for example, through blood transfusion or via contaminated surgical instruments. There would be no 'species barrier' to protect against such transmission. Concern regarding this possibility has been heightened since the demonstration of experimental transmission of BSE between sheep by blood transfusion (Houston et al. 2000; Hunter et al. 2002). Precautionary measures have been put in place in the UK to reduce the risk of such transmission (e.g., leucodepletion of blood and sourcing of pooled plasma products from outside of the UK), but the possibility that secondary transmission may increase the eventual size of the epidemic cannot be excluded.

4 Future Prospects from Epidemiological Studies

Epidemiological surveillance played a key role in the identification of the epidemics of BSE and of vCJD. Continued surveillance to identify all cases of vCJD, both in the UK and elsewhere, is clearly an important priority. In addition, it will be important to remain alert to the possibility that BSE infection of humans may cause disease that is not immediately identifiable as vCJD on clinical or pathological grounds. This is perhaps of special concern with respect to the possibility of disease in those who are other than MM at codon 129 of the *PRNP* gene.

Surveillance and epidemiological studies are impeded currently by the absence of a test for infection with the vCJD agent. This has parallels with trying to study the epidemiology of HIV infection but only being able to observe cases of AIDS, as was the situation in the early 1980s. The development of a serological test for HIV infection not only enabled the epidemiology of the infection to be studied in much greater detail, but also it was possible to target control measures, for example, a safe blood supply, utilizing such diagnostic tests. The tests for infection that are currently available for vCJD generally depend upon being able to sample neural tissue and positive results tend to be obtained only relatively late in the course of infection. The possible early involvement of lymphoid tissue in human infections offers another avenue for testing. Evidence of the infectious agent has been found consistently in tonsillar tissue from vCJD cases (Hill et al. 1999) and such evidence has also been found in stored specimens of appendix tissue taken from vCJD patients 8 months (Hilton et al. 1998) and 2 years prior to disease onset, but not in a sample taken about 9 years before onset (Hilton et al. 2002). On the basis of these findings large-scale unlinked anonymous surveys have been started of stored appendix tissue and fresh tonsil tissue taken from individuals without symptoms of vCJD. Initial results of a retrospective survey of appendix tissue indicated one test-positive among 8318 specimens examined (Hilton et al. 2002). Interpretation of this result is complicated by ignorance of the sensitivity and specificity of the test procedure at different stages in the incubation period. Such surveys also pose ethical dilemmas as identifying 'positive' and 'negative' samples using a test of unknown sensitivity and specificity makes interpretation at the individual level problematic—which is why the tests have been done using unlinked anonymous survey methods.

The availability of a sensitive and specific test that can be used on easily accessible tissue (e.g., blood) very early in the incubation period would be an enormous step forward with respect to advancing understanding of the epidemiology of infection, to determining the likely size of the vCJD epidemic and to aiding the implementation of control measures.

The study of risk factors for vCJD, comparing information on cases with that on controls without the disease, has been relatively uninformative to date with respect to identifying the likely routes of transmission of vCJD. It is highly plausible that this was through the consumption of contaminated beef products, but definitive evidence of this is lacking. However, the case-control approach is an insensitive technique for elucidating dietary associations for common food items, especially as BSE may have contaminated a wide variety of products. Even though such studies have been unrewarding to date, they must continue in the hope of identifying others clues to transmission and to identify iatrogenic transmission if and when it occurs. It is possible that study of the relatively small number of cases of vCJD that have occurred so far outside of the UK may be as or more informative as exposures in other countries may have been less ubiquitous.

References

Almond J, Pattison J. Human BSE. Nature 1997; 389:437–438

Alperovitch A, Will RG. Predicting the size of the vCJD epidemic in France. C.R.Acad. Sci III Biologies 2002; 325:33–36

Alperovitch A, Zerr I, Pocchiari M, Mitrova E, de Pedro Cuesta J, Hegyi I et al. Codon 129 prion protein genotype and sporadic Creutzfeldt-Jakob disease. Lancet 1999; 353:1673–1674

Anderson RM, Donnelly CA, Ferguson NM, Woolhouse MEJ, Watt CJ, Udy HJ et al. Transmission dynamics and epidemiology of BSE in British cattle. Nature 1996; 382:779–788

Andrews NJ, Farrington CP, Cousens SN, Smith PG, Ward HJT, Knight RS et al. Incidence of variant Creutzfeldt-Jakob disease in the UK. Lancet 2000; 356:481–482

Andrews NJ, Farrington CP, Ward HJT, Cousens SN, Smith PG, Molesworth AM et al. Deaths from variant Creutzfeldt-Jakob disease in the UK. Lancet 2003; 361:751–752

Asante EA, Linehan JM, Desbruslais M, Joiner S, Gowland I, Wood AL et al. BSE prions propagate as either variant CJD-like or sporadic CJD-like prion strains in transgenic mice expressing human prion protein. EMBO J 2002; 21:6358–6366

Bateman D, Hilton D, Love S, Zeidler M, Beck J, Collinge J. Sporadic Creutzfeldt-Jakob disease in a 18-year-old in the UK. Lancet 1995; 346:1155–1156

Britton TC, al-Sarraj S, Shaw C, Campbell T, Collinge J. Sporadic Creutzfeldt-Jakob disease in a 16-year-old in the UK. Lancet 1995; 346:1155

Brown P, Preece M, Brandel JP, Sato T, McShane L, Zerr I et al. Iatrogenic Creutzfeldt-Jakob disease at the millennium. Neurology 2000; 55:1075–1081

Bruce ME, Will RG, Ironside JW, McConnell I, Drummond D, Suttie A et al. Transmissions to mice indicate that 'new variant' CJD is caused by the BSE agent. Nature 1997; 389:498–501

Bryant G, Monk PN. Final report of the investigation into the North Leicestershire cluster of variant Creutzfeldt-Jakob., http://www.leics-ha.org.uk/Publics/vcjd_rep.doc, 2001

Cooper JD, Bird SM. UK dietary exposure to BSE in beef mechanically recovered meat: by birth cohort and gender. J Cancer Epidemiol Prevent 2002; 7:59–70

Cousens S, Smith PG, Ward H, Everington D, Knight RSG, Zeidler M et al. Geographical distribution of variant Creutzfeldt-Jakob disease in Great Britain, 1994–2000. Lancet 2001; 357:1002–1007

Cousens SN, Linsell L, Smith PG, Chandrakumar M, Wilesmith JW, Knight RSG et al. Geographical distribution of variant CJD in the UK (excluding Northern Ireland). Lancet 1999; 353:18–21

Cousens SN, Vynnycky E, Zeidler M, Will RG, Smith PG. Predicting the CJD epidemic. Nature 1997a; 385:197–198

Cousens SN, Zeidler M, Esmonde TF, DeSilva R, Wilesmith JW, Smith PG et al. Sporadic Creutzfeldt-Jakob disease in the United Kingdom: analysis of epidemiological surveillance data for 1970–96. Br Med J 1997b; 315:389–395

d'Aignaux JH, Costagliola D, Maccario J, de Villemeur TB, Brandel JP, Deslys JP et al. Incubation period of Creutzfeldt-Jakob disease in human growth hormone recipients in France. American Academy of Neurology 1999; 53:1197–1201

Det Norske Veritas Consulting. Sources of BSE infectivity, http://www.foodstandards.gov.uk/multimedia/pdfs/sources_bse_infect.pdf, 2002
Ferguson NM, Donnelly CA, Ghani AC, Anderson RM. Predicting the size of the epidemic of new variant Creutzfeldt-Jakob disease. Br Food J 1999; 101:86–98.
Ferguson NM, Ghani AC, Donnelly AC, Hagenaars TJ, Anderson RM. Estimating the human health risk from possible BSE infection of the British sheep flock. Nature 2002; 709:1–5
Gail MH, Brookmeyer R. Methods for projecting course of acquired immunodeficiency syndrome epidemic. J Natl Cancer Inst 1988; 80:900–911
Ghani AC, Ferguson NM, Donnelly CA, Hagenaars TJ, Anderson RM. Estimation of the number of people incubating variant CJD. Lancet 1998; 352:1353–1354
Ghani AC, Donnelly CA, Ferguson NM, Anderson RM. Updated projections of future vCJD deaths in the UK. BMC Infect Dis. 2003 3:4
Glatzel M, Rogivue C, Ghani A, Streffer JR, Amsler L, Aguzzi A. Sharply increased Creutzfeldt-Jakob disease mortality in Switzerland. Lancet 2002; 360:139–141
Gore SM. Commentary: Age related exposure of patients to the agent of BSE should not be downplayed. Br Med J 1997; 315:395–396
Gore SM. More than happenstance: Creutzfeldt-Jakob disease in farmers and young adults. Br Med J 1995; 311:1416–1418
Harris-Jones R, Knight R, Will RG, Cousens S, Smith PG, Matthews WB. Creutzfeldt-Jakob disease in England and Wales 1980-84: a case-control study of potential risk factors. J Neurol, Neurosurg Psychiatr 1988; 51:1113–1119
Hill AF, Butterworth RJ, Joiner S, Jackson G, Rossor MN, Thomas DK et al. Investigation of variant Creutzfeldt-Jakob disease and other human prion diseases with tonsil biopsy samples. Lancet 1999; 353:183–189
Hill AF, Desbruslais M, Joiner S, Sidle KCL, Gowland I, Collinge J. The same prion strain causes vCJD and BSE. Nature 1997; 389:448–450
Hilton DA, Fathers E, Edwards P, Ironside JW, Zajicek J. Prion immunoreactivity in appendix before clinical onset of variant Creutzfeldt-Jakob disease. Lancet 1998; 352:703–704
Hilton DA, Ghani AC, Conyers L, Edwards P, McCardle L, Penney M et al. Accumulation of prion protein in tonsil and appendix: review of tissue samples. Br Med J 2002; 325:633–634
Horn G, Bobrow M, Bruce M, Goedert M, McLean A, Webster J. Review of the Origin of BSE: Report to UK Government. 2001 www.defra.gov.uk/animalh/bse/bseorigin.pdf
Houston F, Foster JD, Chong A, Hunter N, Bostock CJ. Transmission of BSE by blood transfusion in sheep. Lancet 2000; 356:999–1000
Huillard d'Aignaux J, Costagliola D, Maccario J, Billette de Villemeur T, Brandel JP, Deslys JP et al. Incubation period of Creutzfeldt-Jakob disease in human growth hormone recipients in France. Neurology 1999; 53:1197–1201
Huillard d'Aignaux JN, Cousens SN, Maccario J, Costagliola D, Alpers MP, Smith PG et al. The incubation period of kuru. Epidemiology 2002; 13:402–408
Huillard d'Aignaux JN, Cousens SN, Smith PG. Predictability of the UK variant Creutzfeldt-Jakob disease epidemic. Science 2001; 294:1729–1731
Hunter N, Foster JD, Chong A, McCutcheon S, Parnham DW, Eaton S et al. Transmission of prion diseases by blood transfusion. J Gen Virol 2002; 83:2897–2905

Jackson GS, Beck JA, Navarrete C, Brown J, Sutton PM, Contreras M et al. HLA-DQ7 antigen and resistance to variant CJD. Nature 2001; 414:269–270

Kao RR, Gravenor MB, Baylis M, Bostock CJ, Chihota CM, Evans JC et al. The potential size and duration of an epidemic of bovine spongiform encephalopathy in British sheep. Science 2002; 295:332–335

Laplanche JL, Lepage V, Peoc'h K, Delasnerie-Laupretre N, Charron D. HLA in French patients with variant Creutzfeldt-Jakob disease. Lancet 2003; 361:531–532

Lee HS, Brown P, Cervenakova L, Garruto RM, Alpers MP, Gajdusek DC et al. Increased susceptibility to Kuru of carriers of the PRNP 129 methionine/methionine genotype. J Infect Dis 2001; 183:192–196

Liberski PP, Gajdusek DC. Kuru: forty years later, a historical note. Brain Pathol 1997; 7:555–560

Lloyd SE, Onwuazor ON, Beck JA, Mallinson G, Farrall M, Targonski P et al. Identification of multiple quantitative trait loci linked to prion disease incubation period in mice. Proc Natl Acad Sci USA 2001; 98:6279–6283

Lowe J. Evidence of a CJD epidemic may still be missed. Br Med J 2000; 320:1011

McLean AR, Bostock CJ. Scrapie infections initiated at varying doses: an analysis of 117 titration experiments. Philos Trans R Soc Lond B Biol Sci 2000; 355:1043–1050

Minor PD, Will RG, Salisbury D. Vaccines and variant CJD. Vaccine 2000; 19:409–410

National CJD Surveillance Unit. Creutzfeldt-Jakob disease surveillance in the UK, http://www.cjd.ed.ac.uk/rep2001.html, 2002

Parry HB. Scrapie disease in sheep: historical, clinical, epidemiological and practical aspects of the natural disease. London, Academic Press, 1983

Pepys MB, Bybee A, Booth DR, Bishop MT, Will RG, Little A-M et al. MHC typing in variant Creutzfeldt-Jakob disease. Lancet 2003; 361:487–489

Prusiner SB, Gajdusek DC, Alpers MP. Kuru with Incubation Periods Exceeding Two Decades. Am Neurolog 1982; 12:1–9

Sartwell PE. The distribution of incubation periods of infectious disease. Am J Hyg 1950; 51:310–318

Valleron A-J, Boelle P-Y, Will R, Cesbron J-Y. Estimation of epidemic size and incubation time based on age characteristics of vCJD in the United Kingdom. Science 2001; 294:1726–1728

van Duijn CM, Delasnerie-Laupetre N, Masullo C, Zerr I, de Silva R, Wientjens DPWM et al. Case-control study of risk factors of Creutzfeldt-Jakob disease in Europe during 1993-95. Lancet 1998; 351:1081–1085

Verity CM, Nicoll A, Will RG, Devereux G, Stellitano L. Variant Creutzfeldt-Jakob disease in UK children: a national surveillance study. Lancet 2000; 356:1224–1227

Wientjens DPWM, Davanipour Z, Hofman A, Kondo K, Matthews WB, Will RG et al. Risk factors for Creutzfeldt-Jakob disease: a reanalysis of case-control studies. Neurology 1996; 46:1287–1291

Wilesmith JW, Ryan JBM, Hueston WD, Hoinville LJ. Bovine spongiform encephalopathy: epidemiological features 1985-1990. Veterinary Record 1992; 130:90–94

Will RG, Ironside JW, Zeidler M, Cousens SN, Estibeiro K, Alperovitch A et al. A new variant of Creutzfeldt-Jakob disease in the UK. Lancet 1996; 347:921–925

Will RG, Matthews WB, Smith PG, Hudson C. A retrospective study of Creutzfeldt-Jakob Disease in England and Wales 1970-1979 II: epidemiology. J Neurol, Neurosurg Psychiatr 1986; 49:749–755

Working Party on Bovine Spongiform Encephalopathy. Report of the Working Party on Bovine Spongiform Encephalopathy (Southwood Report). London, UK, Dept of Health and Ministry of Agriculture, Food and Fisheries, 1989

Chronic Wasting Disease of Cervids

M. W. Miller[1] · E. S. Williams[2]

[1] Colorado Division of Wildlife, Wildlife Research Center, 317 West Prospect Road, Fort Collins, CO 80526-2097, USA
E-mail: mike.miller@state.co.us

[2] Department of Veterinary Sciences, University of Wyoming, 1174 Snowy Range Road, Laramie, WY 82070, USA
E-mail: storm@uwyo.edu

Abstract Chronic wasting disease (CWD) has recently emerged in North America as an important prion disease of captive and free-ranging cervids (species in the deer family). CWD is the only recognized transmissible spongiform encephalopathy (TSE) affecting free-ranging species. Three cervid species, mule deer (*Odocoileus hemionus*), white-tailed deer (*O. virginianus*), and Rocky Mountain elk (*Cervus elaphus nelsoni*), are the only known natural hosts of CWD. Endemic CWD is well established in southern Wyoming and northern Colorado, and has been present in this 'core area' for two decades or more. Apparently CWD has also infected farmed cervids in numerous jurisdictions, and has probably been endemic in North America's farmed deer and elk for well over a decade. Several free-ranging foci distant to the Colorado–Wyoming core area have been discovered since 2000, and new or intensified surveillance may well identify even more foci of infection. Whether all of the identified captive and free-ranging foci are connected via a common original exposure source remains undetermined. Some of this recently observed 'spread' may be attributable to improved detection or natural movements of infected deer and elk, but more distant range extensions are more likely caused by movements of infected captive deer and elk in commerce, or by some yet unidentified exposure risk factor. Research on CWD over the last 5 years has resulted in a more complete understanding of its pathogenesis and epidemiology. CWD is infectious, transmitting horizontally from infected to susceptible cervids. Early accumulation of PrP^{CWD} in alimentary tract-associated lymphoid tissues during incubation suggests agent shedding in feces or saliva as plausible transmission routes. Residual infectivity in contaminated environments also appears to be important in sustaining epidemics. Improved tests allow CWD to be reliably diagnosed long before clinical signs appear. Implications of CWD are not entirely clear at this time. Natural transmission to humans or traditional domestic livestock seems relatively unlikely, but the possibility still evokes public concerns; impacts on wildlife resources have not been determined. Consequently, where CWD is not known to occur surveillance programs and regulations that prevent or reduce the likelihood that CWD will be introduced into these jurisdictions should be encouraged. Where CWD is known to occur, affected jurisdictions are conducting surveillance to estimate and monitor trends in geographic distribution and prevalence, managing deer and elk populations in attempts to limit spread, and developing and evaluating techniques for further controlling and perhaps eradicating CWD. Programs for addressing the challenges of CWD management will require interagency cooperation, commitment of funds and personnel, and applied research.

Chronic wasting disease (CWD) is perhaps the most enigmatic of the naturally occurring prion diseases. Although recognized as a transmissible spongiform encephalopathy (TSE) since the late 1970s (Williams and Young 1980, 1982), interest in and concern about CWD has only recently emerged. CWD most closely resembles scrapie in sheep in most respects, but recent media and public reaction to CWD has been more reminiscent of that afforded to bovine spongiform encephalopathy (BSE) less than a decade ago. Yet, with the exception of transmissible mink encephalopathy (TME), CWD is the rarest of the known animal TSEs: fewer than 1,000 cases have been diagnosed worldwide, and all but two of these occurred in North America. CWD is unique among the TSEs in that it affects free-living species (Spraker et al. 1997; Miller et al. 2000). The three natural host species for CWD, mule deer (*Odocoileus hemionus*), white-tailed deer (*O. virginianus*), and Rocky Mountain elk (*Cervus elaphus nelsoni*), are all in the family Cervidae and native to North America. Like scrapie, CWD is contagious: epidemics are self-sustaining in both captive and free-ranging cervid populations (Miller et al. 1998, 2000). The geographic extent of endemic CWD in free-ranging wildlife was initially thought to be quite limited and its natural rate of expansion slow; however, recent investigations have revealed that CWD has been inadvertently spread much more widely via market-driven movements of infected, farmed elk and deer. Both the ecological and economic consequences of CWD and its spread remain to be determined; moreover, public health implications remain a question of intense interest. Here, we review current understanding of CWD, its implications, and its management.

1
The Emergence of CWD

The precise time and place of most emerging diseases' origins cannot be determined with certainty; CWD is no exception. Data on distribution and occurrence in free-ranging deer and elk suggest that one CWD epidemic may have originated somewhere in northcentral Colorado or southeastern Wyoming, USA (Miller et al. 2000). The original observation of 'chronic wasting disease' was as a clinical syndrome of unknown etiology affecting captive mule deer at wildlife research facilities in northcentral Colorado, in the late 1960s (Williams and Young 1980). Deer in these facilities came from several sources, including free-ranging populations, and interchanges with a wildlife research facility in Wyoming were routine. In 1978, Williams and Young (1980) diagnosed this 'chronic wasting disease' as a form of spongiform encephalopathy by examining brains from symptomatic deer. Shortly thereafter, they diagnosed CWD among captive mule deer in Wyoming (Williams and Young 1980) and in elk from these same facilities (Williams and Young 1982). Two zoological parks, one in the USA and one in Canada, also yielded CWD cases in subsequent years (Williams and Young 1992), but epidemics apparently were not sustained in either of the latter locations.

CWD was recognized for the first time in a free-ranging elk in Colorado in 1981 (Spraker et al. 1997). Within a decade, CWD was diagnosed in free-ranging elk in Wyoming, and in free-ranging mule deer and white-tailed deer in both states (Williams and Miller 2002).

Although this pattern has been interpreted as evidence of CWD's recent emergence, epidemiological data and modeling suggest CWD may have persisted in some of these free-ranging deer populations for >20 years before being detected (Miller et al. 2000). The geographic proximity of affected research facilities and affected free-ranging cervid populations seems to implicate a common origin; however, it is impossible to discern retrospectively whether captive or free-ranging populations represent the original source of CWD.

The potential for CWD's spread via movements of farmed cervids in commerce was recognized several years before it was first detected in this industry (Williams and Young 1992; Miller and Thorne 1993). The first CWD diagnosis in farmed elk was made in Saskatchewan, Canada, in 1996, but epidemiological investigations incriminated South Dakota as the likely source of the Canadian herd's infection; ultimately, the source may have been somewhere in the Wyoming–Colorado endemic area, but no documentation of the latter connection exists. Since then, CWD has been diagnosed in farmed elk facilities from South Dakota (eight cases), Nebraska (four), Oklahoma (one), Colorado (fourteen confirmed), Montana (one), Kansas (one), Minnesota (one), and Wisconsin (one), USA, and in Alberta (one), Canada, as well as in 39 Saskatchewan facilities (Canadian Food Inspection Agency 2002; US Animal Health Association 2001). Infected elk also were exported from Saskatchewan to South Korea in 1997, representing the first known extension of CWD distribution beyond North America. More recently, CWD has been detected on captive white-tailed deer facilities in Wisconsin (three) and Alberta (one). Based on initial epidemiological investigations, it appears likely that CWD has been in the North American farmed cervid industry since at least the late 1980s, and perhaps much longer.

In 2000–2003, cases of CWD in free-ranging deer or elk were diagnosed in westcentral Saskatchewan, in northwest Nebraska, in southwest South Dakota, in southcentral Wisconsin, on the western slope of the Rocky Mountains in northwestern Colorado, in southern New Mexico, in northern Illinois, and in three separate locations in Utah. In the Saskatchewan, Nebraska, and South Dakota cases, CWD-infected game farms probably served as sources of infection for local deer populations. Epidemiological data from Nebraska are especially compelling in suggesting that CWD in free-ranging deer originated in farmed cervids, or at least within the confines of an affected captive cervid facility (Nebraska Game and Parks Commission 2002). The origin of CWD in the northwestern Colorado deer is still under investigation, but index cases were found among free-ranging deer entrapped in an elk farm. In all, publicly owned deer with CWD have been culled from within the confines of elk farms in three states (South Dakota, Nebraska, and Colorado). The origin and extent of CWD foci in Wisconsin, New Mexico, Illinois, and Utah are under active investigation..

2
Causative Agent and Host Range

As with the other known TSEs, CWD appears to be caused by prions, proteinaceous infectious agents devoid of nucleic acids (Prusiner 1982, 1999). CWD behaves as an infectious disease in all experience to date; spontaneous and familial forms have not been identified, but may occur. As with other details of its emergence, the origin of the prion strain that now causes CWD (PrP^{CWD}) is not known. PrP^{CWD} could have resulted from spontaneous conformational alteration of PrP^{c} to PrP^{res} with subsequent transmission to susceptible deer and elk. The common sporadic form of Creutzfeldt–Jakob Disease (CJD) apparently arises by this mechanism (Gajdusek 1996); however, sporadic animal TSEs have not been documented. Alternatively, PrP^{CWD} could have arisen as a cervid-adapted strain of scrapie (Race et al. 2002). The moderate ability of PrP^{Sc} to convert cervid PrP^{C} in vitro (Raymond et al. 2000), PrP^{res} glycoform pattern similarities (Race et al. 2002), the susceptibility of elk to intracerebral exposure to scrapie agent (A. N. Hamir, personal communication), and the susceptibility of goats to intracerebral exposure to the CWD agent (Williams and Young 1992) offer some support for a causal link between scrapie and CWD. An unidentified prion strain also could have been the origin of CWD. Based on mouse strain typing and glycoform pattern comparisons, PrP^{CWD} differs from the BSE agent, some strains of scrapie, and TME agent (Bruce et al. 1997; 2000; Race et al. 2002). The marked similarity of central nervous system lesions, epidemiology, and glycoform patterns strongly suggest that the CWD agent is essentially the same in farmed and free-ranging deer and elk (Williams and Young 1993; Spraker et al. 2002b), but whether multiple strains of PrP^{CWD} occur in nature remains under study.

Mule deer, white-tailed deer, and Rocky Mountain elk are the only species known to be naturally susceptible to CWD. It is likely that subspecies of these three cervid species also are naturally susceptible. A number of wild and domestic species appear to be naturally resistant to CWD, or at least much less susceptible than deer and elk. Moose (*Alces alces*), pronghorn antelope (*Antilocapra americana*), Rocky Mountain bighorn sheep (*Ovis canadensis canadensis*), mouflon (*Ovis musimon*), mountain goats (*Oreamnos americanus*), and a blackbuck (*Antilope cervicapra*) in contact with CWD-affected deer and elk or resident in infected premises have not developed CWD. Domestic livestock appear to be naturally resistant to CWD. A few cattle, sheep, and goats have resided in research facilities with CWD for prolonged periods without developing the disease. Cattle intensively exposed to CWD-infected deer and elk via oral inoculation or confinement with infected captive mule deer and elk have remained healthy for over 5 years (E. S. Williams, M. W. Miller, and T. J. Kreeger, unpublished results), and range cattle in CWD-endemic areas appear to be TSE-free (Gould et al. 2003). These apparent barriers to efficient interspecies transmission are supported by molecular and intracerebral challenge studies (Raymond et al. 2000; Hamir et al. 2001; E. S. Williams and S. Young, unpublished results). As with other TSEs, many species are experimentally susceptible to CWD when exposed via intracerebral inoculation (Williams and Young 1992; Williams and Miller 2002; Williams et al. 2002).

3
Epidemiology

Although CWD is demonstrably both transmissible and infectious, understanding of its natural transmission remains incomplete. CWD epidemics are not sustained by exposure to food-borne infectivity associated with rendered ruminant meat and bonemeal, as are BSE epidemics (Wilesmith et al. 1988); however, the possibility that prion-contaminated feed may initiate a CWD epidemic cannot be discounted. Data from CWD epidemics in captive deer and elk provide strong evidence of lateral transmission (Williams and Young 1992; Miller et al. 1998; Miller and Williams 2003), suggesting that CWD epidemiology is more similar to that of scrapie (Hoinville 1996) than that of BSE. Experimental and epidemic modeling data support these observations (Miller et al. 2000; Gross and Miller 2001; Miller and Williams 2003). Maternal transmission appears to be relatively rare and of negligible importance in maintaining epidemics (Miller et al. 1998, 2000). Some interspecies transmission probably occurs among susceptible host species; there is circumstantial evidence for transmission from mule deer to elk, mule deer to white-tailed deer, and elk to mule deer and white-tailed deer.

The mechanism for agent shedding is perhaps the most perplexing facet of CWD epidemiology. Presumed CWD agent (PrP^{CWD}) accumulates in gut-associated lymphoid tissues (e.g., tonsils Peyer's patches, and mesenteric lymph nodes; Sigurdson et al. 1999; Miller and Williams 2002; Spraker et al. 2002b), thereby supporting alimentary tract (feces and saliva) shedding as the most plausible route. In light of the potential for agent persistence in the environment (Brown and Gajdusek 1991), it seems likely that transmission could be both direct and indirect. Transmission probably requires more than just transient exposure. It follows that minimal fence-line contact probably does not pose excessive risk of transmission, but that prolonged fence-line contact might increase the possibility of transmission. Similarly, concentrating deer and elk in captivity or by artificial baiting or feeding probably increases the likelihood of direct and indirect transmission between individuals. In some epidemics, contaminated pastures appear to have served as sources of infection (Miller et al. 1998; Williams et al. 2002). Similar phenomena have been reported in some scrapie outbreaks (Greig 1940; Pálsson 1979; Andreoletti et al. 2000). Persistence of PrP^{CWD} in contaminated environments may represent a significant potential obstacle to eradication of CWD from either farmed or free-ranging cervid populations. However, CWD apparently did not persist in several facilities that experienced only a few cases and presumably were not heavily contaminated.

The overall duration of CWD infection has been difficult to measure in natural cases (Williams et al. 2002); when during the course of infection an animal may become infectious is equally unclear. Based on data from experimental pathogenesis studies, it appears likely that PrP^{CWD} shedding is progressive through the disease course in deer. The presence of PrP^{CWD} in alimentary tract associated lymphoid tissues early in the incubation period (Sigurdson et al. 1999) suggests that shedding of the agent may begin early. Epidemic models suggest shedding probably precedes onset of clinical disease in both deer and elk (M. W. Miller, unpublished results).

4
Immunity and Natural Resistance

Antibody responses to PrP^{CWD} have not been detected. As in other TSEs, there is some evidence of host response to CWD infections suggested by glial activation in the brain (Hadlow 1996). Similarly, the role of genotype in conferring resistance or increasing susceptibility to CWD remains unclear. In elk, codon 132 methionine homozygotes were overrepresented among CWD-affected individuals when compared with unaffected individuals (O'Rourke et al. 1999), suggesting potential for differential susceptibility. Resistance associated with genotype has not been recognized in deer, but is under active investigation. Although the vast majority of captive deer and elk residing in endemic research facilities eventually contract CWD (Williams and Young 1980; Miller et al. 1998; M. W. Miller, unpublished results; E. S. Williams and T. J. Kreeger, unpublished results), individuals occasionally survive a lifetime in these facilities without succumbing to CWD. Whether genotype or some other unidentified host factor contributes to the likelihood of survival in the face of CWD exposure remains in question.

5
Detection of CWD-Infected Cervids

Detection of CWD in captive and free-ranging cervids is based on various combinations of clinical observation and laboratory diagnostics that form the foundation for established surveillance strategies. These surveillance strategies have been successfully used both in detecting new foci of infection and in estimating prevalence in affected populations.

Clinical observation remains a common tool for detecting CWD in both captive and free-ranging cervids. The most striking and widely recognized clinical features of end-stage CWD in deer and elk are behavioral changes and loss of body condition (Williams and Young 1980, 1982). However, signs of CWD are progressive and early cases of CWD often go unrecognized as such by casual observers. Caretakers or others familiar with individual animals often notice the subtle changes in behavior several months before those unfamiliar with that particular animal can detect abnormalities. Consequently, casual inspections of farmed, captive, or free-ranging cervids may fail to detect evidence of clinical CWD, even when experienced clinicians are used. Affected animals sometimes alter their interaction with conspecifics or handlers. Some show repetitive behaviors (e.g., moving in a set pattern) or periods of somnolence or depression from which they are easily roused. They may carry their heads and ears lowered. Affected animals continue to eat but food consumption declines, contributing to gradual loss of condition. Noticeable weight loss often occurs well after the earliest behavioral changes arise; such losses are more difficult to discern in free-ranging cervids that naturally undergo a seasonal change in body condition.

As CWD progresses, affected animals may display increased drinking and urination, increased salivation with resultant slobbering or drooling, as well as ataxia, subtle trembling, and wide-legged stance. Uncontrollable regurgitation, excitability, and fainting are seen less consistently. In all experience to date, affected animals do not recover from clinical CWD. Aspiration pneumonia, presumably from dysphagia and excess salivation, may confound the diagnosis of CWD if appropriate tissues are not examined. 'Sudden deaths' following handling also have been reported in some situations, as have unusual mortalities (e.g., an elk getting its head caught under a fence). Once infected deer and elk begin to show clinical signs of CWD, they may succumb within a few days or survive for about a year; however, most clinically affected individuals survive less than 4 months. Although the protracted clinical course described above is most common, acute death in deer does occur (M. W. Miller, unpublished results). Signs of CWD tend to be more subtle and the disease course more prolonged in elk than in deer. Overall, clinical courses in free-ranging cervids are probably somewhat shorter than in captivity because compromised abilities to forage, find water, and avoid predators all lessen their chance of survival in the wild.

Because clinical signs are neither consistent nor diagnostic, CWD diagnosis must be confirmed by examination of the brain for spongiform lesions (Williams and Young 1980, 1982, 1993) and/or accumulation of PrP^{CWD} in brain and lymphoid tissues by immunohistochemistry (IHC; Miller et al. 2000; Peters et al. 2000; Miller and Williams 2002; Spraker et al. 2002a). The histopathology of CWD has been described extensively (Williams and Young 1980, 1982, 1993). The parasympathetic vagal nucleus in the dorsal portion of the medulla oblongata at the obex is the most important site to examine for diagnosing CWD (Williams and Young 1993; Peters et al. 2000). This nucleus shows consistent, early involvement following CWD infection in all three known susceptible species. Optimally, the obex should be preserved in 10% buffered formalin and remaining portions of brainstem should be frozen for prospective use in confirmation or strain typing.

Detecting accumulation of PrP^{CWD} in brain and lymphoid tissues via IHC has largely replaced histopathology in diagnosing CWD (Miller et al. 2000; Miller and Williams 2002). Of the several monoclonal antibodies evaluated for CWD diagnosis, 99/97.6.1 (O'Rourke et al. 2000) has proven most reliable (Miller and Williams 2002; Spraker et al. 2002a). IHC of the parasympathetic vagal nucleus at the obex is both sensitive and specific for CWD diagnosis in deer and elk (Miller et al. 2000; Miller and Williams 2002; T. R. Spraker, personal communication). Because PrP^{CWD} accumulates in lymphoid tissues of deer and elk early in infection (Sigurdson et al. 1999; E. S. Williams and M. W. Miller, unpublished results; T. R. Spraker, personal communication), lymphoid tissue IHC also has been used to diagnose CWD. Tonsils or retropharyngeal lymph nodes have proven particularly reliable for diagnosing CWD in deer via IHC, thus providing a foundation for both antemortem and preclinical postmortem diagnostic applications (Miller and Williams 2002; Wolfe et al. 2002). Both tonsils and retropharyngeal lymph nodes are readily collected from heads of harvested animals, and can be examined alone or in combination with brainstem to enhance chances of proper diagnosis.

Laboratory tests developed for BSE (e.g., Deslys et al. 2001) and scrapie are presently being evaluated for use in CWD diagnostics. Preliminary data indicate that some of these tests exhibit reliability similar to IHC (e.g., both sensitive and specific in preclinical cervids), and thus may afford more rapid testing than is possible with IHC. Validations of several such tests for use on brain, tonsils, and lymph nodes in all three susceptible cervid hosts are currently in progress.

Using the foregoing diagnostic tools, several strategies have been devised for surveillance of farmed and free-ranging cervid populations to detect CWD infections. Three basic approaches have emerged, based on laboratory screening of (a) cervids showing clinical signs suggestive of CWD (clinically targeted surveillance), (b) all naturally occurring mortalities, or (c) apparently healthy individuals randomly sampled from populations of interest. Where adequate records or data and individual animal identification or movement information are available, epidemiological investigations can help to identify populations potentially exposed and at risk of infection.

Among farmed cervid herds, CWD has been detected through various combinations of clinically targeted or mortality-based surveillance and epidemiological investigations. Some infected herds have been detected through voluntary or mandated surveillance programs, but surveillance data from most farmed cervid herds are presently insufficient to assure that these herds are not infected with CWD. In farmed herds newly experiencing CWD, sporadic cases of prime-aged animals losing condition, being unresponsive to symptomatic treatment, and dying from pneumonia are commonly reported. Given the variety in initial presentations of clinical CWD, careful laboratory examinations of all juvenile (>6 months old) and adult cervid mortalities on game farms appears prudent and has been required in some jurisdictions. Because incubation periods are uncertain and clinical presentation and course vary, a minimum of 5 years of complete surveillance of all juvenile (>6 months old) and adult mortalities seems the minimum standard necessary to provide relative assurance that farmed cervid herds do not have CWD.

CWD in free-ranging deer and elk has been detected through both clinically targeted surveillance and random sampling. Clinically targeted surveillance is a particularly efficient approach for detecting new foci of infection in free-ranging populations (Miller et al. 2000), and is recommended practice for all North American wildlife management agencies (Williams et al. 2002). Examining road-killed cervids for CWD appears to be somewhat less sensitive than clinically targeted surveillance, but may still be a relatively efficient method for detecting foci of infection (M. W. Miller, unpublished results); road-kill data may tend to overestimate local CWD prevalence, and should be interpreted accordingly. Geographically based random surveys, through IHC screening of either hunter harvested or culled animals, also have been instrumental in detecting CWD foci in Colorado, Wyoming, Saskatchewan, Nebraska, South Dakota, and Wisconsin. Random sampling may be particularly effective when exposure risk (e.g., proximity to CWD-infected farmed cervid facilities, proximity to infected free-ranging populations, proximity to scrapie-infected sheep flocks) can be identified a priori to guide the geographic focus for such surveys. More generalized random sampling of harvested deer and elk also can be used to survey for CWD in free-ranging populations, but appears to be a relatively inefficient method for detecting new disease foci. Moreover, in order for such surveys to be meaningful, sample sizes must be sufficient to detect relatively low infection rates (e.g., 1–2%) with considerable confidence (e.g., ≥95%) in the population of interest. Random surveys provide relatively unbiased estimates of prevalence in infected populations (Conner et al. 2000).

6 Distribution and Occurrence

Understanding of the distribution and occurrence of CWD has changed dramatically over the last decade as diagnostic tools improved and surveillance programs developed and expanded. It now appears that there are at least two contemporary CWD epidemics. One primarily involves free-ranging cervids residing in contiguous portions of Wyoming, Colorado, and Nebraska, USA. The other is more geographically diffuse and primarily involves farmed elk and deer scattered throughout North America. These two epidemics are essentially independent and have limited geographic or causal overlap. The most parsimonious explanation of their origins is through a common root. However, such a root origin has never been reliably demonstrated; if it really exists, then it probably dates back several decades and thus the exact relationships may never be determined. Whether recent discoveries of CWD in captive white-tailed deer and in other free-ranging foci represent new epidemics or simply new facets of one of these two recognized epidemics remains to be determined.

The primary recognized CWD epidemic in free-ranging cervids spans contiguous portions of northern Colorado, southern Wyoming, and western Nebraska, USA; this is often called the 'core CWD endemic area'. Infected deer and elk populations occupy over 60,000 km^2 of native habitats ranging from mountains and foothills in the western portions to river bottoms and shortgrass prairie tablelands in the eastern portions (Miller et al. 2000). Natural geographic spread of the free-ranging endemic focus has been relatively predictable, and patterns appear to coincide with natural movements of deer and elk in affected areas (Miller et al. 2000; M. M. Conner and M. W. Miller, unpublished results). Based on data from ongoing surveillance programs, CWD prevalence varies from less than 1% to more than 15% among the numerous free-ranging cervid populations in this area (Miller et al. 2000; Colorado Division of Wildlife, unpublished results; Wyoming Game and Fish Department, unpublished results). Geographically distinct foci of CWD infection recently reported in free-ranging deer or elk in western Saskatchewan, northwestern Nebraska, southwestern South Dakota, northwestern Colorado, and southern New Mexico probably are not direct extensions of this endemic core. The latter foci all appear to be relatively small and prevalence relatively low, although surveillance and epidemiological investigations are not yet completed in these sites. The extents of the recently recognized foci of CWD in southcentral Wisconsin/northern Illinois and northwestern Colorado are not yet determined, but may be larger than originally believed.

CWD is much more geographically widespread among farmed cervids, and the full extent of its distribution in this industry has yet to be determined. Cases have been diagnosed in numerous states, two Canadian provinces, and at least one location in South Korea, although most infected facilities have been depopulated within several months of being detected. Based on recent trends, surveillance programs will probably uncover new CWD-infected farmed cervid herds in the coming years; the recent discovery of CWD in farmed white-tailed deer is particularly troubling because virtually no surveillance has been conducted in the North American deer industry to date. At present, uniformity and consistency in surveillance for CWD among farmed cervids is lacking, thus precluding a reliable assessment of distribution in this industry. In contrast to the predictable pattern of natural spread in free-ranging cervids, tracking CWD's spread via farmed elk and deer has been highly unpredictable because animal movements are commercial, essentially random, and inadequately regulated or documented in many locations. Undetected spread via trade of infected animals will likely continue until more rigorous and uniform surveillance programs are adopted and enforced. Prevalence among captive cervids can be remarkably high (Williams and Young 1980; Miller et al. 1998); rates of less than 1% to more than 50% have been encountered among infected farmed cervids at depopulation (Nebraska Game and Parks Commission 2002; Peters et al. 2000; Colorado Department of Agriculture, unpublished results).

7 Implications For Public, Livestock, and Wildlife Health

As with other animal TSEs, the public health implications of CWD overshadow more tangible implications for the health of important wildlife resources. Despite media innuendo to the contrary, no cases of human prion disease have been associated with CWD (World Health Organization 2000; Belay et al. 2001; Food and Drug Administration Transmissible Spongiform Encephalopathy Advisory Committee 2001). None of three young people diagnosed with CJD who either hunted or consumed venison were connected epidemiologically to CWD exposure (Belay et al. 2001), and only one of three Wisconsin hunters in a widely reported 'CJD cluster' actually suffered from CJD (E. Belay, personal communication). In vitro studies (Raymond et al. 1998, 2000) support the notion that a natural species barrier probably helps to reduce human susceptibility to CWD and other animal prion diseases. Conversion of human PrP^{C} to the abnormal isoform in the presence of PrP^{CWD} is inefficient compared to homologous PrP^{CWD}–cervid PrP^{C} and PrP^{CJD}–human PrP^{C} conversions; moreover, cervid PrP^{CWD}–human PrP^{C} conversions were essentially equivalent to ovine PrP^{Sc}–human PrP^{C} and PrP^{BSE}–human PrP^{C} conversions (Raymond et al. 2000). This otherwise reassuring molecular evidence of species barriers is clouded by the disparate experiences with scrapie and BSE as human pathogens. Scrapie exposure has not been causally connected to CJD risk despite extensive human exposure to scrapie through handling and consumption of sheep tissues, including brain. In contrast, after intensive exposure (Ghani et al. 2000) of British and other European citizens to the BSE agent, about 120 cases of variant CJD had been confirmed in the UK as of November 2002 (The UK Creutzfeldt–Jakob Disease Surveillance Unit 2002).

Because information on risk is admittedly incomplete, public health officials and wildlife management professionals have undertaken aggressive campaigns to inform the hunting public and others about CWD and where it occurs via the news media, web sites, educational publications, and through specific detailed information in hunting license application packets. For several years, four common-sense measures have been recommended to hunters, meat processors, and taxidermists handling carcasses of cervids that that might be infected with CWD (Colorado Division of Wildlife 2002; Williams and Miller 2002; Williams et al. 2002). These four measures appear sufficient to minimize exposure to CWD as well as other known zoonotic pathogens, and include: (a) avoiding harvest or consumption of sick deer and elk; (b) wearing latex or rubber gloves when dressing a deer or elk from CWD areas; (d) discarding the brain, spinal cord, lymph nodes, spleen, tonsils, and eyes because these organs appear to contain the greatest amount of CWD agent in infected animals; and (d) thoroughly washing knives and other implements used in processing. Sodium hypochlorite (household bleach, >2% free chlorine, equivalent to a 40–50% aqueous solution, at room temperature for 1 h) or sodium hydroxide (caustic soda, soda lye, equivalent to a 4% aqueous solution, at room temperature for 1 h) can be used for disinfection, though these solutions may be corrosive to metal utensils. Deboning game meat also has been recommended as a way to further reduce potential for exposure to the CWD agent. Raw velvet antler, a product unique to the farmed cervid industry, may deserve further evaluation for presence of PrP^{CWD}.

Concerns also have been raised about the possibility that CWD could cross species barriers and infect cattle or sheep sharing grazing areas with infected cervids. CWD transmission to cattle would be particularly devastating for the US and Canadian beef industries. Several studies, both completed and ongoing, suggest that cattle are unlikely to be naturally susceptible to CWD. In vitro conversion of bovine PrP^{C} by PrP^{CWD} was relatively inefficient compared to conversions by PrP^{BSE} or ovine PrP^{Sc} (Raymond et al. 2000). Intracerebral (IC) inoculation of cattle with CWD-infected brain homogenate was similarly inefficient, producing evidence of infection in only five of 13 cattle over a 66-month period (Hamir et al. 2001; J. M. Miller, personal communication); in comparison, all cattle inoculated IC with ovine PrP^{Sc} died 14–18 months post inoculation (PI) (Cutlip et al. 1994). Cattle orally inoculated with CWD brain homogenate have shown no signs of illness at 77 months PI (E.S. Williams and M. W. Miller, unpublished results), and cattle living in paddocks with CWD-infected deer remain healthy after 78 months of exposure (M.W. Miller and E.S. Williams, unpublished results). Finally, histologic and IHC surveillance of brainstems from adult cattle grazing in CWD endemic areas has not revealed any suspect TSE lesions (Gould et al. 2003). In combination, data from these studies provide convincing evidence that CWD does not readily transmit to cattle.

Beyond these perceived risks to human and traditional livestock health, CWD has tangible implications for farmed cervid resources in jurisdictions where it occurs. Farmed cervid herds are invariably quarantined once CWD is diagnosed, thereby limiting uses and value of infected and exposed animals. Infected herds are usually depopulated, but indemnity has been made available only recently in the US; in Canada, depopulating infected farmed elk herds has cost the Canadian government over C$30 million to date in indemnity and clean-up funds (U.S. Animal Health Association 2001; G. Luterbach, personal communication). Guidelines for managing infected farm herds are being developed by animal health officials, and national programs are nearing adoption in both Canada and the US. In the interim, however, lack of comprehensive surveillance or antemortem diagnostics may allow continued spread in commerce, including more incidents of international spread beyond the North American continent. Spillover of CWD into local free-ranging cervid populations apparently has occurred in several locations (Williams et al. 2002); additional spillover incidents could establish more CWD foci, subsequently impacting both cervid farming and wildlife management in those areas.

The implications of CWD for free-ranging deer and elk resources may be even more significant than for farmed cervids. Once identified, CWD-endemic areas are not used as sources for deer and elk translocations. Surveillance programs are needed to assess the extent of infected areas and inform publics about the status of CWD in that area; such programs are expensive and compete for funds needed by other wildlife management programs. Beyond these logistical and administrative complications, the impacts of CWD on deer and elk populations are presently unknown. Epidemic models project patterns of diminished adult survival and population destabilization that could lead to substantial reductions in cervid populations (Gross and Miller 2001). Ultimately, public perceptions and concerns about risks associated with CWD may well erode participation in sport hunting in both CWD-endemic areas and areas where CWD is not known to occur. It follows that CWD could dramatically affect both viability and management of free-ranging cervid herds, as well as other wildlife, in jurisdictions where it occurs.

8
Strategies for Controlling CWD

In light of the serious implications of CWD in both captive and free-ranging cervids, controlling its spread and occurrence seem imperative. Unfortunately, effective strategies for managing infectious prion diseases, including CWD, remain elusive. Long incubation periods, subtle early clinical signs, absence of a practical antemortem diagnostic test, extremely resistant infectious agent, possible environmental contamination, and incomplete understanding of transmission all constrain options for controlling or eradicating CWD. At present, traditional veterinary approaches to therapy and prevention do not apply to CWD and other prion diseases. No treatment is available to prevent infections or recover affected animals; similarly, no vaccine is available to prevent CWD infection in deer or elk. However, both of these areas are presently under active investigation.

In farmed cervid facilities, options for managing CWD are few. These typically include jurisdictional or individual adoption of stringent import requirements to prevent CWD from being introduced, surveillance to detect infected herds, quarantine of infected herds, and depopulation of infected herds. CWD surveillance and herd certification programs and federal programs are nearing adoption in the US and Canada; other countries are in various stages of contemplating their preferred approaches to preventing or disclosing CWD in their farmed cervid industries. In the interim, some jurisdictions have imposed bans on importing or moving cervids until the full extent of the CWD problem in farmed cervids is better understood.

The few attempts at eradicating CWD have thus far failed. The cause(s) of failures in two research facilities were not determined, but residual environmental contamination following depopulation and facility clean-up was suspected in both cases (Williams and Young 1992; Miller et al. 1998). CWD was not perpetuated in several zoological parks where a few cases were diagnosed, but the basis for these apparent successes remains unclear (Williams et al. 2002). Attempts to eliminate CWD from farmed elk populations are too recent to allow assessment. Whether heavily contaminated environments can ever be completely disinfected remains a subject of debate; sentinel studies have been proposed and are needed to evaluate the efficacy of such practices. Until effective cleaning and disinfection procedures are identified (or environmental contamination is discounted as a potential source of reinfection), farmed cervids should not be reintroduced into facilities where CWD has occurred and free-ranging cervids should be excluded from previously infected premises (Williams and Miller 2002; Williams et al. 2002). The foregoing problems associated with managing infected herds and premises underscore the need for preventive management, including aggressive surveillance and movement restrictions, to prevent infected animals from moving in commerce.

Options for managing CWD in free-ranging deer and elk are even more limited than for captive cervids (Williams et al. 2002). Surveillance programs to estimate and monitor CWD distribution and prevalence have been undertaken by wildlife management agencies in several jurisdictions; in northeastern Colorado, ongoing surveillance is designed to evaluate both natural temporal changes in prevalence and effects of recent management interventions. Management programs established in recent years have focused on containing CWD and reducing its prevalence or eliminating it altogether in localized areas (Miller and Kahn 1999; Colorado Division of Wildlife 2001). Long-term CWD management goals vary among affected jurisdictions. In areas where CWD may not yet be well established or widely distributed (e.g., Saskatchewan, Wisconsin) eradication appears to be the ultimate goal for CWD management. In contrast, wildlife managers in Colorado and Wyoming have refrained from committing to eradication because it appears unattainable in their endemic CWD situations (Colorado Division of Wildlife 2001); recent evidence of wider distribution of CWD along the common border of these two states makes eradication an even less realistic goal.

Where CWD management in free-ranging wildlife is being attempted, a variety of specific strategies have emerged. Translocating cervids from endemic areas has been banned by policy to help limit range expansion. Artificial feeding has been outlawed, or at least discouraged, to minimize concentrations of deer and elk and hopefully decrease transmission. Culling of clinical CWD suspects has been practiced in Colorado and Wyoming for over a decade, but has proven insufficient to reduce prevalence or limit spread in the absence of other management. More recently, apparently healthy deer in the vicinity of CWD cases have been culled to provide data on the extent of a local problem in several jurisdictions. Localized population reductions in areas of high CWD prevalence have been undertaken in Colorado, but efficacy remains to be determined. Density reduction has been advocated as a tool for controlling CWD, but well-established migration patterns and social behaviors typical of deer and elk may diminish the efficacy of wholesale density reduction in reducing or eliminating CWD from free-ranging cervid populations. According to model projections, aggressive intervention by selective culling early in the course of a CWD epidemic should afford the best hope of precluding establishment of endemic foci (Gross and Miller 2001). Because practical limits of surveillance probably hamper detection of new foci in their first decade after introduction (Miller et al. 2000), however, such intervention may be unachievable in practice. Attempts to aggressively reduce deer numbers in newly identified foci have been undertaken in Nebraska, Saskatchewan, Colorado, and Wisconsin; whether CWD can be eliminated from these areas remains to be determined. Availability of tonsil biopsy as a nonlethal means of testing deer for CWD may augment management programs under some conditions where other measures are not possible. Although large-scale programs for testing free-ranging populations seem impractical (Wolfe et al.2002), test-and-cull is being evaluated experimentally as a CWD management strategy in localized areas of northern Colorado (L. L. Wolfe and M. W. Miller, unpublished results).

Despite practical challenges and lack of effective management tools, responsible wildlife management and animal health agencies should continue their attempts to limit distribution and occurrence of CWD in free-ranging and farmed cervids worldwide. Both affected and unaffected jurisdictions should continue exchanging information on CWD epidemiology, as well as working toward a more complete understanding of its ecology and control.

Acknowledgements. We thank B. Morrison, Nebraska; R. Fowler, South Dakota; R. Lind, Saskatchewan; and J. Langenberg, Wisconsin for access to unpublished results from CWD surveillance programs in their respective states and province. K. O'Rourke, E. Belay, L. Creekmore, and A. Hamir also provided helpful communications about unpublished works. Support for research and management of CWD in Colorado and Wyoming in cervids has been provided by the Colorado Division of Wildlife, Wyoming Game and Fish Department, and University of Wyoming, and by Federal Aid in Wildlife Restoration projects in both states.

References

Andreoletti, O., P. Berthon, D. Marc, P. Sarradin, J. Grosclaude, L. van Keulen, F. Schelcher, J. M. Elsen, F. Lantier. 2000. Early accumulation of PrP(Sc) in gut-associated lymphoid and nervous tissues of susceptible sheep from a Romanov flock with natural scrapie. J Gen Virol 81:3115–3126

Belay, E. D., P. Gambetti, L. B. Schonberger, P. Parchi, D. R. Lyon, S. Capellari, J. H. McQuiston, K. Bradley, G. Dowdle, J. M. Crutcher, C. R. Nichols. 2001. Creutzfeldt-Jakob disease in unusually young patients who consumed venison. Arch Neurol 58:1673–1678

Brown, P., D. C. Gajdusek. 1991. Survival of scrapie virus after 3 years' interment. Lancet 337:269–270

Bruce, M. E., A. Chree, E. S. Williams, H. Fraser. 2000. Perivascular PrP amyloid in the brains of mice infected with chronic wasting disease. Brain Pathol 10:662–663

Bruce, M. E., R. G. Will, J. W. Ironside, I. McConnell, D. Drummond, A. Suttie, L. McCardle, A. Chree, J. Hope, C. Birkett, S. Cousens, H. Fraser, C. J. Bostock. 1997. Transmissions to mice indicate that 'new variant' CJD is caused by the BSE agent. Nature 389:498–501

Canadian Food Inspection Agency [CFIA]. 2002. Chronic wasting disease (CWD) of deer and elk. Canadian Food Inspection Agency Fact Sheet. http://www.cfia-acia.agr.ca/english/anima/heasan/disemala/cwdmdce.shtml [Accessed 15 April 2002]

Colorado Division of Wildlife (CDOW). 2002. Chronic wasting disease. Colorado Division of Wildlife. http://wildlife.state.co.us/hunt/huntereducation/chronic.asp [Accessed 15 November 2002]

Colorado Division of Wildlife. 2001. Colorado Wildlife Commission policy: CWD final policy. (Revised June 10 2002) http://wildlife.state.co.us/hunt/huntereducation/CWDfinalpolicy.asp [Accessed 15 November 2002]

Conner, M. M., C. W. McCarty, M. W. Miller. 2000. Detection of bias in harvest-based estimates of chronic wasting disease prevalence in mule deer. J Wildlife Dis 36:691–699

Cutlip, R. C., J. M. Miller, R. E. Race, A. L. Jenny, J. B. Katz, H. D. Lehmkuhl, B. M. DeBey, M. M. Robinson. 1994. Intracerebral transmission of scrapie to cattle. J Infect Dis 169:814–820

Deslys, J. P., E. Comoy, S. Hawkins, S. Simon, H. Schimmel, G. Wells, J. Grassi, J. Moynagh. 2001. Screening slaughtered cattle for BSE. Nature 409:476–478

Food and Drug Administration Transmissible Spongiform Encephalopathy Advisory Committee. 2001. Transcripts of open meeting. 18 January, Bethesda, Maryland, USA. www.fda.gov/ohrms/dockets/ac/01/transcripts/3681t2_02.pdf [Accessed 9 February 2002]

Gajdusek, D. C. 1996. The potential risk to humans of amyloids in animals. Pages 1–7 *in* C. J. Gibbs, Jr., editor. Bovine spongiform encephalopathy: the BSE dilemma. Springer-Verlag, New York, USA

Ghani, A. C., N. M. Ferguson, C. A. Donnelly, R. M. Anderson. 2000. Predicted vCJD mortality in Great Britain. Nature 406:583–584

Greig, J. R. 1940. Scrapie: Observations on the transmission of the disease by mediate contact. Vet J 96:203–206

Gross, J. E., M. W. Miller. 2001. Chronic wasting disease in mule deer: disease dynamics and control. J Wildlife Manage 65:205–215

Gould, D. H., J. L. Voss, M. W. Miller, A. M. Bachand, B. A. Cummings, A. A. Frank. 2003. Survey of cattle in northeast Colorado for evidence of chronic wasting disease: Geographical and high risk targeted sample. J Vet Diagn Invest: 15:274–277

Hadlow, W. J. 1996. Differing neurohistologic images of scrapie, transmissible spongiform encephalopathy, and chronic wasting disease of mule deer and elk. Pages 122–137 *in* C. J. Gibbs, Jr. editor. Bovine spongiform encephalopathy: the BSE dilemma. Springer-Verlag, New York, USA

Hamir, A. N., R. C. Cutlip, J. M. Miller, E. S. Williams, M. J. Stack, M. W. Miller, K. I. O'Rourke, M. J. Chaplin. 2001. Preliminary findings on the experimental transmission of chronic wasting disease agent of mule deer to cattle. J Vet Diagn Invest 13:91–96

Hoinville, L. J. 1996. A review of the epidemiology of scrapie in sheep. Revue Scientifique et Technique 15:827–852

Miller, M. W., R. Kahn. 1999. Chronic wasting disease in Colorado deer and elk: recommendations for statewide monitoring and experimental management planning. Unpublished report, Colorado Division of Wildlife, Denver, USA

Miller, M. W., E. T. Thorne. 1993. Captive cervids as potential sources of disease for North America's wild cervid populations: avenues, implications, and preventive management. Transactions of the North American Wildlife and Natural Resources Conference 58:460–467

Miller, M. W., M. A. Wild, E. S. Williams. 1998. Epidemiology of chronic wasting disease in Rocky Mountain elk. J Wildlife Dis 34:532–538

Miller, M. W., E. S. Williams. 2002. Detecting PrP^{CWD} in mule deer by immunohistochemistry of lymphoid tissues. Vet Rec: 151:610–612

Miller, M. W., E. S. Williams. 2003. Horizontal prion transmission in mule deer. Nature 425:35–36

Miller, M. W., E. S. Williams. C. W. McCarty, T. R. Spraker, T. J. Kreeger, C. T. Larsen, E. T. Thorne. 2000. Epizootiology of chronic wasting disease in free-ranging cervids in Colorado and Wyoming. J Wildlife Dis 36:676–690

Nebraska Game and Parks Commission. 2002. CWD test results, northern Sioux County, NE. Sampling periods: November 2001 and January/February 2002. http://www.ngpc.state.ne.us/furdocs/CWDmaps.html [Accessed 15 April 2002]

O'Rourke, K. I., T. E. Besser, M. W. Miller, T. F. Cline, T. R. Spraker, A. L. Jenny, M. A. Wild, G. L. Zebarth, E. S. Williams. 1999. PrP genotypes of captive and free-ranging Rocky Mountain elk (*Cervus elaphus nelsoni*) with chronic wasting disease. J Gen Virol 80:2765–2769

O'Rourke, K. I., T. V. Baszler, T. E. Besser, J. M. Miller, R. C. Cutlip, G. A. H. Wells, S. J. Ryder, S. M. Parish, A. N. Hamir, N. E. Cockett, A. Jenny, D. P. Knowles. 2000. Preclinical diagnosis of scrapie by immunohistochemistry of third eyelid lymphoid tissue. J Clin Microbiol 38:3254–3259

Pálsson, P. A. 1979. Rida (scrapie) in Iceland and its epidemiology. In S. B. Prusiner W. J. Hadlow, editors. Slow transmissible diseases of the nervous system. Academic Press, New York, USA: 357–366

Peters, J., J. M. Miller, A. L. Jenny, T. L. Peterson, K. P. Carmichael. 2000. Immunohistochemical diagnosis of chronic wasting disease in preclinically affected elk from a captive herd. J Vet Diagn Invest 12:579–582

Prusiner, S. B. 1982. Novel proteinaceous infectious particles cause scrapie. Science 216:136–144

Prusiner, S. B. 1999. Development of the prion concept. In S. B. Prusiner, editor. Prion biology and diseases. Cold Spring Harbor Laboratory Press, Cold Spring Harbor, New York, USA: 67–112

Race, R. E., A. Raines, T. G. M .Baron, M. W. Miller, A. Jenny, E. S. Williams. 2002. Comparison of abnormal prion protein glycoform patterns from transmissible spongiform encephalopathy agent-infected deer, elk, sheep, and cattle. J Virol 76:12365–12368

Raymond, G. J., A. Bossers, L. D. Raymond, K. I. O'Rourke, L. E. McHolland, P. K. Bryant, III, M. W. Miller, E. S. Williams, M. Smits, B. Caughey. 2000. Evidence of a molecular barrier limiting susceptibility of humans, cattle and sheep to chronic wasting disease. EMBO Journal 19:4425–4430

Raymond, G. J., J. Hope, D. A. Kocisko, S. A. Priola, L. D. Raymond, A. Bossers, J. Ironside, R. G. Will, S. G. Chen, R. B. Petersen, P. Gambetti, R. Rubenstein, M. A. Smits, P. T. Lansbury, Jr., B. Caughey. 1997. Molecular assessment of the potential transmissibilities of BSE and scrapie to humans. Nature 388:285–288

Sigurdson, C. J., E. S. Williams, M. W. Miller, T. R. Spraker, K. I. O'Rourke, E. A. Hoover. 1999. Oral transmission and early lymphoid tropism of chronic wasting disease PrPres in mule deer fawns (*Odocoileus hemionus*). J Gen Virol 80:2757–2764

Spraker, T. R., M. W. Miller, E. S. Williams, D. M. Getzy, W. J. Adrian, G. G. Schoonveld, R. A. Spowart, K. I. O'Rourke, J. M. Miller, P. A. Merz. 1997. Spongiform encephalopathy in free-ranging mule deer (*Odocoileus hemionus*), white-tailed deer (*Odocoileus virginianus*), and Rocky Mountain elk (*Cervus elaphus nelsoni*) in northcentral Colorado. J Wildlife Dis 33:1–6

Spraker, T. R., K. I. O'Rourke, A. Balachandran, R. R. Zink, B. A. Cummings, M. W. Miller, B. E. Powers. 2002a. Validation of monoclonal antibody F99/97.6.1 for immunohistochemical staining of brain and tonsil in mule deer (*Odocoileus hemionus*) with chronic wasting disease. J Vet Diagn Invest 14:3–7

Spraker, T. R., R. R. Zink, B. A. Cummings, M. A. Wild, M. W. Miller, K. I. O'Rourke. 2002*b*. Comparison of histological lesions and immunohistochemical staining of proteinase resistant prion protein in a naturally-occurring spongiform encephalopathy of free-ranging mule deer (*Odocoileus hemionus*) with those of chronic wasting disease of captive mule deer. Vet Pathol 39:110–119

The UK Creutzfeldt-Jakob Disease Surveillance Unit. 2002. CJD Statistics. University of Edinburgh, Edinburgh, United Kingdom. http://www.cjd.ed.ac.uk/ [Accessed on 15 November 2002]

U.S. Animal Health Association. 2001. Report of the Committee on Wildlife Diseases. Accessed 15 April 2002 at http://www.usaha.org/reports/reports01/r01wd.html

U.S. Department of Agriculture. 2002. Chronic wasting disease program. Accessed on 14 April 2002 at http://www.aphis.usda.gov/vs/cwd_program.htm

Williams, E. S., M. W. Miller. 2002. Chronic wasting disease in deer and elk in North America. Revue Scientifique et Technique 21:305–316

Williams, E. S., M. W. Miller, T. J. Kreeger, R. H. Kahn, E. T. Thorne. 2002. Chronic wasting disease of deer and elk: A review with recommendations for managment. J Wildlife Manage 66:551–563

Williams, E. S., S. Young. 1980. Chronic wasting disease of captive mule deer: a spongiform encephalopathy. Journal of Wildlife Diseases 16:89–98

Williams, E. S., S. Young. 1982. Spongiform encephalopathy of Rocky Mountain elk. J Wildlife Dis 18:465–471

Williams, E. S., S. Young. 1992. Spongiform encephalopathies of Cervidae. Revue Scientifique et Technique 11:551–567

Williams, E. S., S. Young. 1993. Neuropathology of chronic wasting disease of mule deer (*Odocoileus hemionus*) and elk (*Cervus elaphus nelsoni*). Vet Pathol 30:36–45

Wolfe, L. L., M. M. Conner, T. H. Baker, V. J. Dreitz, K. P. Burnham, E. S. Williams, N. T. Hobbs, M. W. Miller. 2002. Evaluation of antemortem sampling to estimate chronic wasting disease prevalence in free-ranging mule deer. J Wildlife Manage 66:564–573

World Health Organization. 2000. WHO Consultation on public health and animal transmissible spongiform encephalopathies: epidemiology, risk and research requirements. Document WHO/CDS/CSR/APH/2000.2, World Health Organization, Geneva, Switzerland. http://www.who.int/emc-documents/tse/whocdscsraph2002c.html

Subject Index

Current Topics in Microbiology and Immunology

Volumes published since 1989 (and still available)

Vol. 240: **Hammond, John; McGarvey, Peter; Yusibov, Vidadi (Eds.):** Plant Biotechnology. 1999. 12 figs. XII, 196 pp. ISBN 3-540-65104-7

Vol. 241: **Westblom, Tore U.; Czinn, Steven J.; Nedrud, John G. (Eds.):** Gastroduodenal Disease and Helicobacter pylori. 1999. 35 figs. XI, 313 pp. ISBN 3-540-65084-9

Vol. 242: **Hagedorn, Curt H.; Rice, Charles M. (Eds.):** The Hepatitis C Viruses. 2000. 47 figs. IX, 379 pp. ISBN 3-540-65358-9

Vol. 243: **Famulok, Michael; Winnacker, Ernst-L.; Wong, Chi-Huey (Eds.):** Combinatorial Chemistry in Biology. 1999. 48 figs. IX, 189 pp. ISBN 3-540-65704-5

Vol. 244: **Daëron, Marc; Vivier, Eric (Eds.):** Immunoreceptor Tyrosine-Based Inhibition Motifs. 1999. 20 figs. VIII, 179 pp. ISBN 3-540-65789-4

Vol. 245/I: **Justement, Louis B.; Siminovitch, Katherine A. (Eds.):** Signal Transduction and the Coordination of B Lymphocyte Development and Function I. 2000. 22 figs. XVI, 274 pp. ISBN 3-540-66002-X

Vol. 245/II: **Justement, Louis B.; Siminovitch, Katherine A. (Eds.):** Signal Transduction on the Coordination of B Lymphocyte Development and Function II. 2000. 13 figs. XV, 172 pp. ISBN 3-540-66003-8

Vol. 246: **Melchers, Fritz; Potter, Michael (Eds.):** Mechanisms of B Cell Neoplasia 1998. 1999. 111 figs. XXIX, 415 pp. ISBN 3-540-65759-2

Vol. 247: **Wagner, Hermann (Ed.):** Immunobiology of Bacterial CpG-DNA. 2000. 34 figs. IX, 246 pp. ISBN 3-540-66400-9

Vol. 248: **du Pasquier, Louis; Litman, Gary W. (Eds.):** Origin and Evolution of the Vertebrate Immune System. 2000. 81 figs. IX, 324 pp. ISBN 3-540-66414-9

Vol. 249: **Jones, Peter A.; Vogt, Peter K. (Eds.):** DNA Methylation and Cancer. 2000. 16 figs. IX, 169 pp. ISBN 3-540-66608-7

Vol. 250: **Aktories, Klaus; Wilkins, Tracy, D. (Eds.):** Clostridium difficile. 2000. 20 figs. IX, 143 pp. ISBN 3-540-67291-5

Vol. 251: **Melchers, Fritz (Ed.):** Lymphoid Organogenesis. 2000. 62 figs. XII, 215 pp. ISBN 3-540-67569-8

Vol. 252: **Potter, Michael; Melchers, Fritz (Eds.):** B1 Lymphocytes in B Cell Neoplasia. 2000. XIII, 326 pp. ISBN 3-540-67567-1

Vol. 253: **Gosztonyi, Georg (Ed.):** The Mechanisms of Neuronal Damage in Virus Infections of the Nervous System. 2001. approx. XVI, 270 pp. ISBN 3-540-67617-1

Vol. 254: **Privalsky, Martin L. (Ed.):** Transcriptional Corepressors. 2001. 25 figs. XIV, 190 pp. ISBN 3-540-67569-8

Vol. 255: **Hirai, Kanji (Ed.):** Marek's Disease. 2001. 22 figs. XII, 294 pp. ISBN 3-540-67798-4

Vol. 256: **Schmaljohn, Connie S.; Nichol, Stuart T. (Eds.):** Hantaviruses. 2001, 24 figs. XI, 196 pp. ISBN 3-540-41045-7

Vol. 257: **van der Goot, Gisou (Ed.):** Pore-Forming Toxins, 2001. 19 figs. IX, 166 pp. ISBN 3-540-41386-3

Vol. 258: **Takada, Kenzo (Ed.):** Epstein-Barr Virus and Human Cancer. 2001. 38 figs. IX, 233 pp. ISBN 3-540-41506-8

Vol. 259: **Hauber, Joachim, Vogt, Peter K. (Eds.):** Nuclear Export of Viral RNAs. 2001. 19 figs. IX, 131 pp. ISBN 3-540-41278-6

Vol. 260: **Burton, Didier R. (Ed.):** Antibodies in Viral Infection. 2001. 51 figs. IX, 309 pp. ISBN 3-540-41611-0

Vol. 261: **Trono, Didier (Ed.):** Lentiviral Vectors. 2002. 32 figs. X, 258 pp. ISBN 3-540-42190-4

Vol. 262: **Oldstone, Michael B.A. (Ed.):** Arenaviruses I. 2002, 30 figs. XVIII, 197 pp. ISBN 3-540-42244-7

Vol. 263: **Oldstone, Michael B. A. (Ed.):** Arenaviruses II. 2002, 49 figs. XVIII, 268 pp. ISBN 3-540-42705-8

Vol. 264/I: **Hacker, Jörg; Kaper, James B. (Eds.):** Pathogenicity Islands and the Evolution of Microbes. 2002. 34 figs. XVIII, 232 pp. ISBN 3-540-42681-7

Vol. 264/II: **Hacker, Jörg; Kaper, James B. (Eds.):** Pathogenicity Islands and the Evolution of Microbes. 2002. 24 figs. XVIII, 228 pp. ISBN 3-540-42682-5

Vol. 265: **Dietzschold, Bernhard; Richt, Jürgen A. (Eds.):** Protective and Pathological Immune Responses in the CNS. 2002. 21 figs. X, 278 pp. ISBN 3-540-42668-X

Vol. 266: **Cooper, Koproski (Eds.):** The Interface Between Innate and Acquired Immunity, 2002, 15 figs. XIV, 116 pp. ISBN 3-540-42894-1

Vol. 267: **Mackenzie, John S.; Barrett, Alan D. T.; Deubel, Vincent (Eds.):** Japanese Encephalitis and West Nile Viruses. 2002. 66 figs. X, 418 pp. ISBN 3-540-42783-X

Vol. 268: **Zwickl, Peter; Baumeister, Wolfgang (Eds.):** The Proteasome-Ubiquitin Protein Degradation Pathway. 2002, 17 figs. X, 213 pp. ISBN 3-540-43096-2

Vol. 269: **Koszinowski, Ulrich H.; Hengel, Hartmut (Eds.):** Viral Proteins Counteracting Host Defenses. 2002, 47 figs. XII, 325 pp. ISBN 3-540-43261-2

Vol. 270: **Beutler, Bruce; Wagner, Hermann (Eds.):** Toll-Like Receptor Family Members and Their Ligands. 2002, 31 figs. X, 192 pp. ISBN 3-540-43560-3

Vol. 271: **Koehler, Theresa M. (Ed.):** Anthrax. 2002, 14 figs. X, 169 pp. ISBN 3-540-43497-6

Vol. 272: **Doerfler, Walter; Böhm, Petra (Eds.):** Adenoviruses: Model and Vectors in Virus-Host Interactions. Virion and Structure, Viral Replication, Host Cell Interactions. 2003, 63 figs., approx. 280 pp. ISBN 3-540-00154-9

Vol. 273: **Doerfler, Walter; Böhm, Petra (Eds.):** Adenoviruses: Model and Vectors in Virus-Host Interactions. Immune System, Oncogenesis, Gene Therapy. 2004, 35 figs., approx. 280 pp. ISBN 3-540-06851-1

Vol. 274: **Workman, Jerry L. (Ed.):** Protein Complexes that Modify Chromatin. 2003, 38 figs., XII, 296 pp. ISBN 3-540-44208-1

Vol. 275: **Fan, Hung (Ed.):** Jaagsiekte Sheep Retrovirus and Lung Cancer. 2003, 63 figs., XII, 252 pp. ISBN 3-540-44096-3

Vol. 276: **Steinkasserer, Alexander (Ed.):** Dendritic Cells and Virus Infection. 2003, 24 figs., X, 296 pp. ISBN 3-540-44290-1

Vol. 277: **Rethwilm, Axel (Ed.):** Foamy Viruses. 2003, 40 figs., X, 214 pp. ISBN 3-540-44388-6

Vol. 278: **Salomon, Daniel R.; Wilson, Carolyn (Eds.):** Xenotransplantation. 2003, 22 figs., IX, 254 pp.ISBN 3-540-00210-3

Vol. 279: **Thomas, George; Sabatini, David; Hall, Michael N. (Eds.):** TOR. 2004, 49 figs., X, 364 pp.ISBN 3-540-00534-X

Vol. 280: **Heber-Katz, Ellen (Ed.):** Regeneration: Stem Cells and Beyond. 2004, 42 figs., XII, 194 pp.ISBN 3-540-02238-4

Vol. 281: **Young, John A. T. (Ed.):** Cellular Factors Involved in Early Steps of Retroviral Replication. 2003, 21 figs., IX, 240 pp. ISBN 3-540-00844-6

Vol. 282: **Stenmark, Harald (Ed.):** Phosphoinositides in Subcellular Targeting and Enzyme Activation. 2003, 20 figs., X, 210 pp. ISBN 3-540-00950-7

Vol. 283: **Kawaoka, Yoshihiro (Ed.):** Biology of Negative Strand RNA Viruses: The Power of Reverse Genetics. 2004, 24 figs., approx. 350 pp. ISBN 3-540-40661-1